The Molecular Basis of
Mutation

The Molecular Basis of Mutation

JOHN W. DRAKE

University of Illinois

HOLDEN-DAY

San Francisco, London, Cambridge, Amsterdam

To Juliet and Jonathan

Acknowledgments

This book could not have been prepared without the constant help and encouragement of my wife Pamela.

Trink, Matt, and Frank have also contributed in many indirect but important ways.

It is a pleasure to thank Chuck Beiger for transforming my sketches into all of the drawings in this book and to thank Dona Lindstrom for diligently weeding out many inaccurate and confusing tangles in the text. The remaining deficiencies are of course entirely my own. I am grateful to June Harris and Sue Forsberg for their extensive help with the preparation of the manuscript.

Unpublished results from my own laboratory which are described here were obtained from research supported by the American Cancer Society, the U.S. Public Health Service (Institute of Allergy and Infectious Diseases), and the National Science Foundation.

I would also like to thank the many persons who have allowed me to quote their unpublished results and the authors and publishers who have allowed me to reproduce various versions of previously published figures.

Preface

As a result of the precipitous advances of the past decade in the area of molecular genetics, we now possess a moderately detailed understanding of the mechanical aspects of chromosome replication and segregation, of recombination, and of gene transcription and translation. Studies on mutational processes have played an important dual role in the growth of molecular genetics. Analyses of mutations have frequently provided understanding of other genetic processes, of regulatory and polar mutations in the analysis of gene regulation, for instance, and of frameshift mutations in the analysis of the translation of the genetic code. At the same time, the chemical properties of mutagens have frequently provided crucial insights into mechanisms of mutational processes themselves.

This book aims to present a broad outline of what is understood about mutational mechanisms and also to emphasize many of the doubtful areas. The straightforward biochemical analysis of mutational processes is proving to be an extraordinarily difficult task. The degree of our understanding in the face of this fact represents a considerable triumph for genetics, one which should encourage future workers to fully utilize the methods of genetics in the analysis of other difficult problems (such as those arising in neurobiology). I hope also to stimulate further research into mutational mechanisms themselves, because while the taxonomy of point mutations appears to be well established, it is clear that the experimental evidence supporting many currently popular hypotheses concerning mutational processes is quite inadequate.

Most of the early conceptual and analytical advances in molecular genetics arose from studies on procaryotic microorganisms, particularly bacteriophages. Although more recent years have seen an initial extension of these results into the realm of eucaryotic and multicellular organisms, I have largely avoided discussing mutation in the higher forms, since very little is in fact known about the molecular mechanisms involved. Mutation in eucaryotes will become increasingly subject to analysis, however, and may well involve processes which are qualitatively different from those observed in procaryotes.

Urbana, Illinois
April, 1969

John W. Drake

Contents

1. Prospects for a molecular description of mutation

Ever since mutation was emphasized by de Vries (1901) as a fundamental genetic process, its analysis has occupied a position close to the center of the geneticist's arena. Mutation was quickly recognized as the cause of modified genes which constitute the raw materials for evolution and which, in a myriad of combinations achieved by genetic recombination, are subject to natural selection. Mutation has generally been recognized as such a crucial process that modern definitions of life include mutability as a fundamental property of living organisms. The acquisition of mutant individuals soon became crucial to the study of basic genetic processes in the laboratory and, often, to the production of commercially superior strains of plants and animals. However, for many years the mutational process was described solely in terms of alterations in the morphology or behavior of multicellular organisms.

During the rather extended period when chromosomes were being identified in detail as the physical counterparts of heredity, it became obvious that mutations could be described in a much more satisfactory manner: they could be identified, in the form of mutant genes, as highly localized alterations of chromosomes. In addition, different physical configurations could be recognized among mutations: some behaved as simple points on a chromosome, whereas others behaved in a complex fashion, as if they were comprised of extended alterations of chromosome structure. The more complex mutations could often be correlated cytologically with gross changes in chromosomes, such as losses of recognizable parts, or exchanges of material within and between chromosomes. But, without an understanding of the chemical nature of the chromosome and the gene, these observations did not suggest specific models of mutational mechanisms subject to experimental analysis.

GENETICISTS AND CHEMISTS

The first of several major advances in the analysis of mutational processes was the discovery by Muller in 1927 that mutations could be induced in *Drosophila* by using X irradiation. This was eventually broadened to include mutagenesis by chemical agents; the first success in this direction was achieved in 1941 by Auerbach and Robson (reported in 1947), who induced truly frightening frequencies of mutations in *Drosophila* with mustard gas. Since that time, chemical mutagenesis has frequently been achieved by employing both exotic compounds which may be produced synthetically, and common biochemicals included in the average contemporary human diet. However, the low chemical specificity of most of these reagents, as well as a continuing ignorance of the chemical basis of heredity, greatly delayed the application of these discoveries in determining the chemical basis of the mutation process.

The next major advance was the discovery of the essential role of the nucleic acids in heredity, first demonstrated by Avery, MacLeod, and McCarty (1944) in their studies of the transformation of bacteria by highly purified samples of DNA, then by Hershey and Chase (1952) in their demonstration that the infective moeity of a bacteriophage consists primarily of DNA, and finally by Gierer and Schramm (1956) in their demonstration of the infectivity of the RNA of tobacco mosaic virus. It then became quite obvious (to the farsighted) that the central problem of genetics had become a chemical problem: the determination of the structure and coding properties of DNA.

The final and crucial advance, therefore, was the analysis by Wilkins, Crick and Watson of the structure of DNA (Wilkins et al., 1953; Crick and Watson, 1953). The immediate result of their analyses was the duplex model of DNA which is described in Figures 1-1, 1-2, 1-3, and 1-4. The model, which has survived in good health for a decade and a half, immediately offered insights into the processes of chromosomal replication and mutation (Watson and Crick, 1953 a, b). Many of these ideas were tested and more or less confirmed in their broad outlines within the next few years. Recent studies of the structure of mutational lesions and the events which generate them invariably involve a detailed consideration of the structure of DNA and its component bases.

Although a variety of problems concerning the DNA structure proposed by Watson and Crick are fascinating to organic and physical chemists, the geneticists initially focused their attention on only a few aspects of the structure. The first of these was the redundancy of information within the structure, expressed in complementary polynucleotides running in opposite directions, like a comparative text containing Hebrew on one page and Greek on the opposite. It was therefore to be anticipated that the different strands of the DNA molecule might sometimes express their information in different ways and that primary mutational lesions might lead to internal contradictions within the double helical structure. Secondly, since the four bases are organized into only two diads, while each diad may exist in two configurations, i.e., adenine:thymine and thymine:adenine, it was anticipated that some genetic processes would distinguish only two elements in a code, whereas others would distinguish four. Finally, it seemed obvious that the proposed scheme for the semi-conservative duplication of DNA depended solely upon interactions between bases on opposite strands at a given level and not upon additional species of molecules.

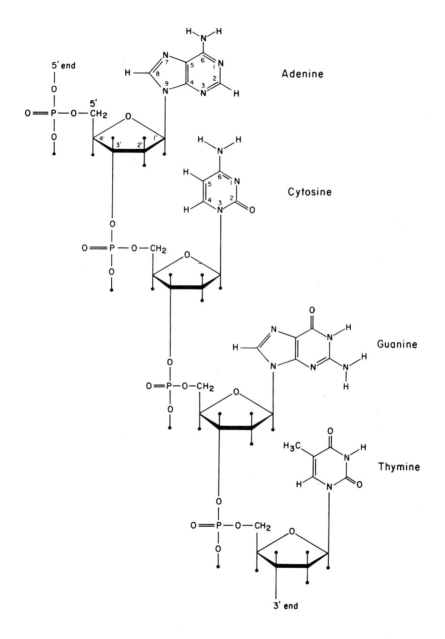

FIGURE 1-1. *The structural components of DNA. Hydrogen atoms are represented by small, solid dots. The polarity of polynucleotides is* 3' → 5'.

THYMINE   ADENINE

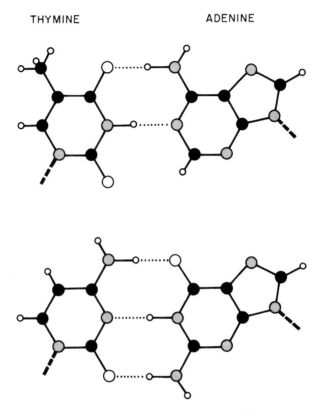

CYTOSINE   GUANINE

FIGURE 1-2. *The standard base pairing arrangements. Solid circles represent carbon atoms; shaded circles are nitrogen atoms; large open circles are oxygen atoms; and small open circles are hydrogen atoms. Solid lines represent covalent bonds; dotted lines are hydrogen bonds; and heavy dashed lines are glycosidic (base-sugar) bonds.*

Further analyses and additional observations, however, strongly indicated that base pairing alone was insufficient to produce the observed levels of accuracy characteristic of DNA replication.

 Modern students of mutation have strongly emphasized three interrelated problems. The first of these has been to deduce the configuration of a mutated region within the DNA molecule, which of course means to determine the original and mutated base pair sequence. The second problem involves tracing the often complicated series of events by which the mutant DNA configuration arose, in particular understanding both the primary chemical event and the fate of what is often called the "premutational lesion." The third problem has been to unravel the often even more complicated effects of the mutation upon the organism, which means tracing the consequences of an altered piece of

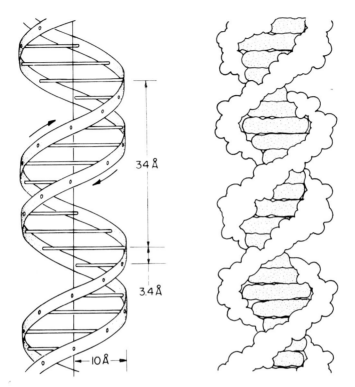

FIGURE 1-3. *The double helix. The schematic model on the left illustrates the opposing polarities of the complementary strands and the characteristic dimensions of the B (wet) form of the molecule. The figure on the right is redrawn from the space-filling model of Feughelman et al. (1955) to illustrate the dense packing of the base pairs in the interior of the structure (shaded) and the "deep" and "narrow" grooves between the deoxyribose phosphate back-bones (open).*

information through the thicket of information transfer and cellular metabolism.

Viruses exhibit all of the properties which have made microorganisms in general such suitable materials for investigations of fundamental biological mechanisms: relative simplicity of composition, genetic homogeneity, haploidy, rapid mode of growth, and the opportunity for examining rare events within huge populations. Specifically, much of our understanding of mutational processes derives from studies of coliform bacteriophages. The genetic and physiological analysis of bacteriophage systems has consistently occupied a central position in the turbulent but ever more successful marriage between genetics and chemistry. At least during the period of courtship, the seductive partner has usually been genetics, particularly by exhibiting the basic problems. However, the chemical analysis of the underlying mechanisms is rapidly catching up and now occasionally even points the way to new genetic studies, although

FIGURE 1-4.   *An early structural model of DNA, constructed by George Gamow and partly composed of personal possessions of J. D. Watson, who also appears.*

the inherent rarity of mutational events has made their chemical analysis forebodingly difficult. In these pages the reader will find a preponderance of genetic studies. Hopefully, an occasional biochemist will be stimulated by the possibilities for investigation which are so richly distributed throughout the area of mutagenesis.

* * * * *

The study of mutational mechanisms has been rather closely interwoven with the study of the genetic code, and in fact some knowledge of the latter subject will be very helpful when reading about mutation. Thorough discussions are available in Woese (1967a, b) and in the many papers in the *Cold Spring Harbor Symposia of Quantitative Biology*, Vol. 31. A number of review articles have also appeared in recent years which summarize diverse aspects of mutation in viruses (Krieg, 1963a; Freese, 1963; Orgel, 1965). Excellent discussions of the history of mutational studies within the context of classical Mendelian genetics are to be found in Sturtevant's *History of Genetics* (1965), in Dunn's *Short History of Genetics* (1965), and in Carlson's *The Gene: A Critical History* (1966). A description of some of the methods and pitfalls of mutational studies in higher organisms appears in Auerbach's monograph *Mutation* (1962).

2. Why bacteriophages?

The motives which led the founders of molecular biology to settle upon bacteriophages as favorable organisms, are certainly a suitable topic for the history of science (Cairns, Stent and Watson, 1966). Although these motives were probably very complex, the choices which were made have been richly rewarded by the growth of the science of bacterial viruses. The problem some three decades ago was to choose simple organisms which would be receptive to the physical and chemical probes then in existence or still to be developed. Crystallization of the tobacco mosaic virus in 1935 had already dramatized the fundamental simplicity and regularity of virus particles. Furthermore, the bacteriophages exhibited a number of major advantages over other viruses: their host cells were easy to cultivate, their assay was simple and rapid, and their pathogenicity was negligible. A fortuitous advantage later became obvious: the specific infectivity of bacteriophages is generally high and often approaches unity, in contrast to infectivity-to-particle ratios from 10^{-2} to 10^{-6} in other virus systems.

It is one of the pleasant accidents of science that the set of bacteriophages initially chosen for intensive study by Max Delbrück and his co-workers turned out to be relatively complicated organisms. Many bacteriophages are so small and uncomplicated that they are unsuitable for the study of genetic processes, such as mutation and recombination (however interesting their other properties may be). The chosen bacteriophages were all rather large particles, ranging from 50 to 300 million daltons in mass, possessing a variety of structural proteins, and, in the case of the largest particles, carrying genetic information sufficient to specify well over a hundred polypeptides. These viruses have turned out to possess DNA of conventional double-stranded structure, although frequently exhibiting other unusual chemical modifications. They also all mutate and recombine at easily measured frequencies.

Some familiarity with viral titration, structure, and reproduction will be necessary for an understanding of the topics which are to follow. A sketch of the fundamentals will be presented in this and in the following chapter. Further details can be easily obtained from *Molecular Biology of Bacterial Viruses* by Stent (1963) and from *General Virology* by Luria and Darnell (1967).

COUNTING BACTERIOPHAGES

Titration of a bacteriophage suspension usually involves making a viable count by the **plaque assay**. A petri dish is filled to a depth of about 5 mm with a solid bottom layer containing nutrient broth, salts, and about 1.2 per cent agar. A top-agar-layer tube containing about 2.5 ml of the same medium, but with only about 0.6 per cent agar, is held in the liquid state at 45°. Just before use, about 10^8 host cells are added to this tube, along with a small predetermined volume of the bacteriophage suspension to be assayed. The contents of the tube are then mixed, poured over the bottom agar layer, allowed to harden, and incubated (usually at 37°) for 6 to 24 hours. Each virus particle initiates a spreading focus of infection on the plate. At the same time, the large number of bacteria originally plated grow into a continuous lawn. The foci of infection appear as holes in this lawn and are called plaques. Since plaques are initiated by single virus particles, their number is a direct count of viable particles. The detailed morphology of plaques is often complex and reflects the genetic composition of the initiating virus. Figure 5-7 shows a variety of plaque types.

The total number of virus particles, viable or inactive, can be conveniently counted in the electron microscope. The particles are suspended in a solution of volatile salts to which is added a known concentration of polystyrene latex particles of about 200 mm diameter. The mixture is then sprayed on grids, dried, and examined. The ratio of virus particles to latex particles in a number of clearly defined droplet patterns provides the desired measure of virus particle concentration.

STRUCTURAL CONSIDERATIONS

The two viruses which have been used most successfully in mutational studies are the "large" bacteriophage T4 and the "small" bacteriophage S13, both of which grow well on various strains of *Escherichia coli.*

Bacteriophage T4 (Figure 2-1) has a mass of about 300×10^6 daltons and is composed of protein, of a single molecule of double-stranded DNA of mass 120×10^6 daltons, and of small amounts of polyamines. (The polyamines appear to neutralize negative charges on the DNA. Since they can be replaced by magnesium ions without affecting the infectivity of the virus particles, they are not very interesting in the present context.) Although a variety of different proteins are present, 85-90 per cent of the protein of the particle consists of the **head membrane** which encloses the viral DNA; this protein is arranged in subunits of molecular weight of about 80,000. The **tail** of bacteriophage T4 is a complex organelle. It is joined to the head by a **collar** of unknown function. Along most of its length the tail consists of two concentric tubular structures, the outer **sheath** and the inner **core**. At the end of the tail is the **tail plate** to which are attached six long **tail fibers** in an outer hexagonal array and six shorter fibers more centrally situated; the short fibers often appear as **spikes** in the free

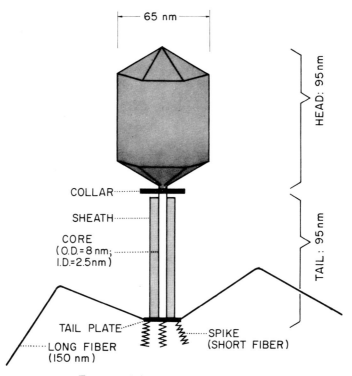

FIGURE 2-1. *Bacteriophage T4.*

particle, but may be observed in their extended position in particles attached to the bacterial surface. Each viral organelle is composed of a number of subunits, which are not necessarily identical.

The tail constitutes the offensive armament of the bacteriophage particle. When a susceptible cell is encountered as a result of random diffusion, the long tail fibers make a firm and host-specific attachment to the bacterial surface. (In the case of bacteriophage T4, for instance, the attachment is made to a specific surface lipopolysaccharide.) The inner fibers then make contact, bringing the end plate into a position close to the cell surface. The sheath contracts, forcing the core through the cell wall and against or through the cell membrane. The DNA then passes out of the head of the bacteriophage, down through a hole in the center of the core, and into the cell. A small amount of protein of unknown function accompanies the DNA into the cell.

The DNA of bacteriophage T4 possesses the double-stranded configuration but exhibits peculiarities which have proved invaluable aids to the analysis of intracellular viral development. Cytosine is completely replaced by **5-hydroxy-methylcytosine** (5HMC). In addition, the 5HMC residues are **glucosylated** after incorporation into phage DNA (Figure 2-2). The appropriate enzymatic apparatus to produce these modifications is coded in the phage DNA. The

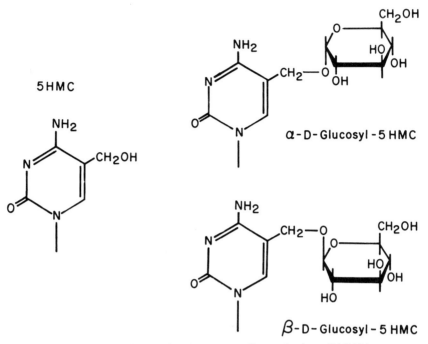

5HMC

α-D-Glucosyl-5HMC

β-D-Glucosyl-5HMC

FIGURE 2-2.   *The novel components of bacteriophage T4 DNA.*

glucosylation of the phage DNA appears to serve a defensive function against cellular nucleases: unglucosylated phage DNA is degraded in most strains of *Escherichia coli* (Richardson, 1966). On the other hand, glucosylation does not function in an hereditary capacity; when T4 multiplies in *Shigella dysenteriae,* it is neither glucosylated nor degraded.

Viewed in the electron microscope, the DNA of bacteriophage T4 consists of an uninterrupted linear structure some 52 nm in length. Microbial geneticists usually broaden the meaning of the term chromosome to cover this structure. However, the apparently simple shape of the molecule is deceptive. If a population of T2 chromosomes (closely related to bacteriophage T4) is heated sufficiently to melt out the hydrogen bonds and separate the component single strands and if the population of single strands is then slowly cooled, complementary strands reanneal with a very high efficiency. The reannealed molecules, however, now appear as circles (Thomas, 1966a). The explanation of this result is that while the various T2 chromosomes each contain the same set of genes arranged in the same order, they are related to each other as if each had once been a circle, and the circles had been opened by randomly distributed breaks. Such chromosomes are said to be **circular permutations** of each other (Figure 2-3).

If a population of T2 DNA molecules is treated very briefly with exonuclease III, an enzyme which digests only the 3′ ends of the molecule and thus leaves

abcdefghijklmnopqrstuvwxyzabc

lmnopqrstuvwxyzabcdefghijklmn

FIGURE 2-3. *The arrangement of genes along the T4 chromosome, illustrating circular permutation (top versus bottom line) and terminal redundancy (beginning versus end of each line).*

single-stranded 5′ ends exposed, the treated molecules show a strong spontaneous tendency to form circles, without the intermediate melting process (Thomas, 1966a). The two ends of the molecule are therefore identical in base pair sequence, and the chromosome is said to be **terminally redundant** (Figure 2-3). The redundancy is of variable length, extending over 0.5 percent to 3 percent of the chromosome and encompassing between about one and perhaps four or five genes.

A model has been proposed by Streisinger (Streisinger, Edgar, and Denhardt, 1964; Séchaud et al., 1966; Streisinger, Emrich, and Stahl, 1967) to account for the existence both of circular permutation and of terminal redundancy in T4. A terminally redundant chromosome first replicates within a host cell. The daughter chromosomes then become joined end to end by an act of recombination, producing a concatenate. By iteration of this process, a very long concatenate may be constructed. When maturation begins, a free end of the concatenate is seized and begins to be packed into a head structure. After a specific *length* (or *volume*) of DNA has been packaged, which is generally slightly longer than one complete set of bacteriophage genes, the concatenate is cut, and the new end is used to start filling another head. (Under the "headful" hypothesis, it is the chromosome length and not its base pair sequence which is of critical importance.) This process is outlined in Figure 2-4. A careful examination of the figure will reveal that, if each headful contains slightly more than one complete set of genes, succeeding chromosomes cut from the concatenate will comprise circular permutations of each other.

The DNA of bacteriophages such as T4 is packed extremely tightly within the head in a denatured (Tikchonenko et al., 1966; Gorin et al., 1967) or dehydrated (Maestre and Tinoco, 1967) state. Since the chemical reactivity of the bases, especially of the amino groups, is sharply reduced when the bases are involved in typical hydrogen bonding, it is not surprising to find that the susceptibility of T4 to certain mutagens is intermediate between the susceptibilities of double-stranded and single-stranded DNA molecules in buffer (Freese and Strack, 1962). Unfortunately, it is not at all clear whether rates of formation of different types of mutational lesions are equally sensitive to the physical state of the target DNA. There is a clear need for a comparative study of mutation within a given gene naked in buffer, packaged within a bacteriophage particle, and resident within cellular cytoplasm.

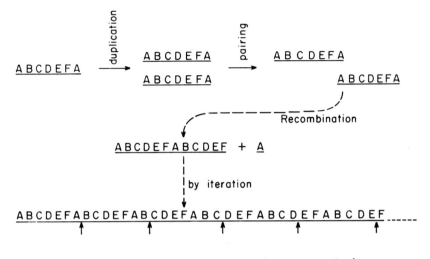

(Arrows indicate cuts produced during maturation)

FIGURE 2-4.   *A proposed mechanism for generating circular permutations and terminal redundancy in the T4 chromosome.*

The structure of the minute bacteriophage S13 and its very close relative ΦX174 contrasts strikingly with that of T4. The particle mass is only 6.2×10^6 daltons, which includes 1.7×10^6 daltons of DNA. The particle is very simple, consisting of a thick protein shell within which the DNA is embedded. The viral DNA is both single-stranded and circular. The small size of the ΦX174 chromosome limits it to about seven genes. The replication of ΦX174 DNA proceeds through a double stranded intermediate which is also circular.

INTRACELLULAR REPRODUCTION

The life cycle of the great majority of virulent bacteriophages is formally described by the the **one-step growth curve**. The virus particles are first synchronously adsorbed to host cells, either by using very high cell concentrations to speed adsorption, or by reversibly poisoning the cells with cyanide during the adsorption period. Development is then allowed to proceed under well defined conditions of nutrition, temperature, and cell density, and samples are taken at regular intervals for plaque assay. An unchanging concentration of plaque-forming units is observed for the first several minutes; this simply represents infected cells. An abrupt increase in the plaque titer of the suspension eventually appears, and soon reaches a plateau value which is constant so long as additional growth cycles are avoided. The cycle is shown graphically as curve A in Figure 2-5, which is plotted semilogarithmically to accommodate the large changes which are observed.

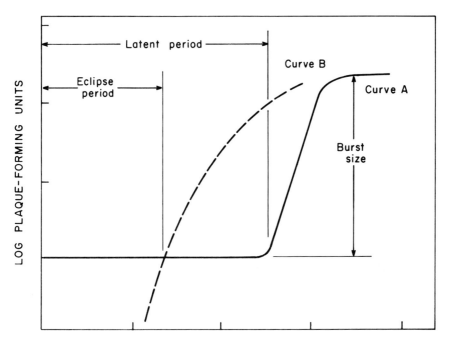

LOG PLAQUE-FORMING UNITS

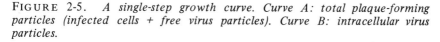

MINUTES AT 37°C

FIGURE 2-5. *A single-step growth curve. Curve A: total plaque-forming particles (infected cells + free virus particles). Curve B: intracellular virus particles.*

The interval preceding the increase in titer is called the **latent period**. The interval of increasing titer is called the rise period or the **release period**; with the great majority of viruses, the release of particles from infected cells is achieved by the lysis of the cell. The ratio of titers before and after lysis is called the **burst size**. The burst size is usually highly variable, depending upon cultural conditions; in addition, the individual cells in a given culture release widely varying numbers of particles. In the more complex viruses such as T4, the latent period tends to be more constant in diverse nutritional environments, although it depends strongly upon temperature. In more simple viruses like ΦX174, even the latent period is subject to modification by nutritional conditions and cell density.

Infected cells may be prematurely broken open, for instance with lysozyme and chloroform, and assayed for their content of bacteriophages. (Most bacteriophages are very resistant to chloroform, and a drop or two is often placed in the stock vial to maintain sterility.) The plaque-forming ability of the bacteriophage particle disappears immediately after infection, a result not surprising in view of the process of adsorption and penetration which has been described. The first newly synthesized virus particles make their intracellular appearance after approximately one-half of the latent period. The interval between infection and the first intracellular appearance of new virus particles (when an average of one virus particle is to be found per cell) is called the **eclipse period**. After the end of the eclipse period, virus particles accumulate in large numbers until released by lysis. The intracellular synthesis of virus particles is traced by curve B in Figure 2-5.

Viral components are synthesized in a well-ordered sequence. From some seven genes (for ΦX174) to well over a hundred (for T4) must act during the latent period. The first to act are those which are concerned with the inactivation of the cellular genome and the establishment of enzymatic systems for the synthesis of viral DNA. Consider, for example, the case of T4 growing at 37°. The enzymes required for the synthesis of 5HMC begin to be synthesized within two minutes after infection. The class of **early enzymes** can often be further subdivided according to the times when their synthesis is initiated and halted. At a later time (about 6 minutes), viral DNA synthesis begins. Very shortly thereafter, viral somatic protein synthesis begins. (A somatic protein is one which appears in the mature particle.) This is followed (at 10-12 minutes) by the intracellular appearance of the first mature particles. While the rate of DNA synthesis initially increases rapidly, it soon reaches a constant value; this value is just matched by the rate at which viral DNA is encapsulated into maturing particles. The result is a steady-state pool of viral DNA, the vegetative pool, within which events such as genetic recombination and replication-dependent mutation are localized. The size of this pool depends upon cultural conditions, and may range from as few as 15 to as many as 60 **phage equivalents**. (A phage equivalent is the amount of any substance present in a single bacteriophage particle.) The sequence of events outlined above is depicted in Figure 2-6.

The mechanism which so well orders bacteriophage development remains a major mystery. The operons which frequently constitute the units of control in bacteria seem to be largely absent from bacteriophages, although both T4 (Stahl et al., 1966; Nakata and Stahl, 1967) and S13 (Tessman et al., 1967) contain single small units of coordinate transcription. The shutoff of early enzyme synthesis is closely linked to the initiation of viral DNA synthesis (see Epstein et al., 1963), but the nature of the linkage is obscure.

LYSOGENY

Avoidance of massive damage to the host is often a cardinal virtue of parasites. Many bacteriophages exhibit a more generous mode of interaction with host cells than the purely lytic cycle which has been described above. Upon infection, and with a probability less than unity, certain viruses may establish a lysogenic state. In the instances which have been well characterized, this is achieved by the following series of events. The viral DNA, upon entering the cell, assumes a circular configuration by joining together its free ends. (In the case of

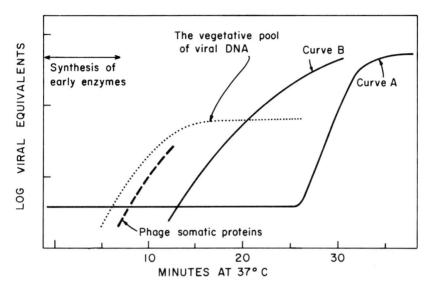

FIGURE 2-6. *The timing of synthetic processes in viral development; curves A and B are the same as in Figure 2-5, but refer specifically to bacteriophage T4.*

bacteriophage λ, for instance, the free ends have short, single-stranded regions some 15 bases long which are complementary, and which simply anneal.) By means of a single break-reunion recombination event (see Chapter 3), the circle integrates into the host chromosome. (The virus and host share a common nucleotide sequence which permits this exchange.) Thereafter, the viral DNA replicates as a part of the host chromosome, and is said to be in the **prophage** state; a virus capable of this form of interaction with its host is said to be **temperate**. The functioning of the viral genes is repressed, apparently just as certain host genes are frequently repressed. At any later time, either spontaneously or in response to external conditions, the process may be reversed, and may lead to a normal lytic cycle initiated from within the cell. Until then, however, the lysogenic cell is immune to infection by the strain of the virus which it harbors. Most strains of bacteria probably harbor one or more prophages.

3. Bacteriophage genetics: first principles

The elements of virus genetics basic to an understanding of mechanisms of virus mutation will be briefly developed here. In particular, we will consider the replication of bacteriophage DNA, the recognition of genes by means of their mutant alleles, the process of genetic recombination, and finally, certain aspects of recombination and gene expression peculiar to viruses. Genetic mapping will be considered in Chapter 4. Most of the present discussion will refer to bacteriophage T4 as a model system, but where relevant, other viruses such as ΦX174 will also be considered.

DNA REPLICATION

Only DNA and a small amount of "internal protein" enter the bacterium upon infection by bacteriophage T4. If the parental DNA is density-labelled, for instance with 5-bromouracil or C^{13} and N^{15}, it is observed to achieve a hybrid density starting some 5-6 minutes after infection (Kozinski et al., 1963; Roller, 1964). Replication is therefore semi-conservative. However, because of the intervention of genetic recombination (to be discussed below), T4 hybrid DNA molecules are extensively broken and recombined, often with totally light (granddaughter) molecules; the result is that only a short, single-stranded piece of the original DNA molecule generally appears in any given progeny molecule (Figure 3-1). This complication is much less evident when a bacteriophage such as λ is examined, whose rate of genetic recombination is markedly lower than that of T4 (Meselson and Weigle, 1961). Furthermore, if several λ particles infect the same cell, a small fraction of them may escape replication altogether, and emerge repackaged in a new protein coat. The study of such "free riders" may be useful to differentiate mutational events which depend upon DNA replication, from mutational events which depend only upon tenure in cytoplasm.

Bacteriophage T4 encodes its own DNA polymerase, whose gene (prosaically called "gene 43") exhibits extremely interesting properties, which will be

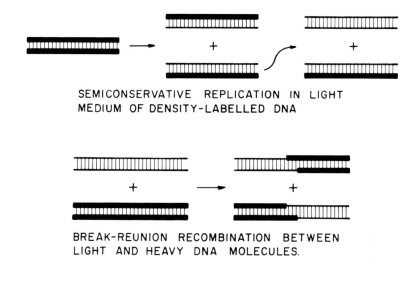

SEMICONSERVATIVE REPLICATION IN LIGHT
MEDIUM OF DENSITY-LABELLED DNA

BREAK-REUNION RECOMBINATION BETWEEN
LIGHT AND HEAVY DNA MOLECULES.

A TYPICAL REPLICATED, RECOMBINANT,
DENSITY-LABELLED T4 DNA MOLECULE

FIGURE 3-1. *DNA replication (top) and genetic recombination (middle) in bacteriophage T4 traced by the distribution of density-labeled parential DNA. The two processes acting simultaneously produce chromosomes of the type shown at the bottom.*

considered in Chapter 15. The cellular DNA polymerase alone is unable to replicate T4 DNA. Replicating T4 DNA is tied up in a protein complex, and appears to be extensively "nicked" (that is, carrying single-strand interruptions), perhaps as a result of its involvement in recombination (Kozinski and Lin, 1965). In addition, as will be discussed below, it may achieve a length much longer than that of a single chromosome.

When bacteriophage ΦX174 infects a cell, its (circular) DNA is first converted to a double-stranded form. This function appears to be carried out by a cellular, and not a viral, polymerase. The two subsequent modes of replication, semi-conservative replication of the double-stranded form and synthesis of much larger numbers of single-stranded molecules, appear to be under viral control (Tessman, 1966; Tessman et al., 1967; Denhardt and Sinsheimer, 1965b), but the enzymology of these steps remains to be elucidated. It is also likely that the double-stranded molecule which contains the initially infecting viral strand remains in a special category and functions differently from its descendents (Denhardt and Sinsheimer, 1965b, c).

DISCOVERING GENES

A gene is recognized by means of a mutant version of itself. The existence of a gene nowadays frequently is inferred from the existence of a specific protein or a specific catalytic process, without actual possession of the appropriate mutants. However, its existence under these conditions remains only an inference and not an identification.

The range of mutants which appear among bacteriophages will be discussed in Chapter 7; only a few examples need be mentioned here. Under standard and very carefully controlled conditions, plaques exhibit reproducible shapes and sizes. Among several thousand T4 plaques derived from a common ancestor, a few appear with distinctly larger, smaller, or more turbid appearance. Since most of these variants breed true, they are mutants. Examples of large-plaque (r) mutants appear in Figure 3-5. Cells resistant to a virus may easily be selected, and extended host range (h) mutants capable of infecting these resistant cells may in turn be selected.

The symbolism of viral genetics is partially implied in these two examples; the rubric r, for instance, derives from *rapid lysis*. The mutant gene is usually symbolized by italicised lower case symbols. The corresponding unmutated, or wild type gene is signified by adding a "+" superscript: r^+, h^+. When mutational alteration of any of several different genes results in mutants of the same general appearance, the different genes are usually specified by additions to the initial symbol. For instance, a number of different genes are involved in maintaining the r^+ condition in T4 and r mutants may arise as a result of damages within any of them. As these genes were first differentiated, they were called rI, rII, and rIII; even more are now known, but they have not been named in print. Further investigations indicated that the rII gene in fact consisted of a pair of adjacent genes, which were then called rIIA and rIIB.

To further complicate matters, large numbers of mutants may be obtained which affect the same gene, and which may or may not have distinctive properties (such as stability, extent of inactivation of the gene, location within the gene, and molecular geometry). In most laboratories, these mutants are differentiated by serial numbers, often simply representing the order in which the mutants were isolated. Capitalized letters may be included in order to distinguish different sets of mutants, representing, for instance, the individual who performed the isolations, or the mutagen employed to induce the mutants. Thus, within the rIIB cistron, the mutant rUV375 was induced by ultraviolet irradiation, while the mutant rFC11 arose spontaneously and was isolated in the laboratory of Francis Crick.

Since mutants of similar appearance do not necessarily contain modifications of the same gene, further criteria are required to decide whether different mutant individuals are affected in different genes. The appropriate measurement is called a **complementation test**. In this test, two mutant chromosomes are introduced into a common cytoplasm, and the system is then observed to determine whether the two chromosomes complement each other, that is, whether the wild type condition is restored. The method by which different chromosomes may be inserted into a common cytoplasm will obviously depend upon the characteristics of the organism; for forms exhibiting sexuality, it is only necessary to examine the diploid zygote, providing it is sufficiently stable. For bacteriophages, multiple infection of the host cell usually suffices. Similarly, the determination of whether complementation has occurred depends upon the character in question. Since plaque morphology is expressed long after single infection of a cell on the plate, it is not suitable for measuring complementation between *r* mutants; instead, an appropriately modified single-step growth experiment must be performed.

As a specific example of a complementation test, consider two independently arising mutants defined by their inability to grow at 42°. (Most bacteriophages grow fairly well at 42°. Of necessity, most *ts* mutants grow fairly well at some lower temperature, frequently 30°.) Bacteria are infected with an average of about five particles of each mutant to ensure that all cells are infected with both mutants, and then are incubated at 42° throughout the latent period. After lysis, the average burst size is measured. If the two mutational lesions reside in different genes, the burst size usually will be nearly normal: each chromosome provides the gene which is damaged in the other chromosome. If the two lesions

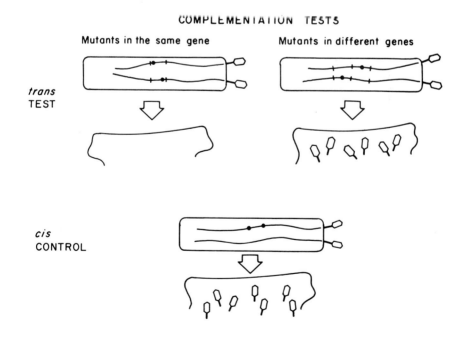

FIGURE 3-2.   *Complementation tests in a bacteriophage system.*

reside in the same gene, the burst size will be very small. A logically necessary control, which in practice is frequently omitted, is to test the combination of a wild type chromosome and a doubly mutant chromosome. (The latter is constructed by recombination between the two single mutants; see below.) The burst size of the control should be normal.

Several factors may complicate complementation tests. Mutants in different genes sometimes fail to complement, or else complement weakly, whereas even mutants in the same gene will sometimes complement each other weakly. Examples of misleading complementation tests will be considered in detail at various points in later chapters; the general problems they pose cannot be treated in depth at this time.

A viral complementation test is illustrated schematically in Figure 3-2. In the test, the two mutant lesions occupy *trans* positions, whereas in the control they occupy *cis* positions. As a result, genes which are defined by means of *cis-trans* complementation tests (Lewis, 1951) were called **cistrons** by Benzer (1957). The word has since come into more common use as a substitute for *gene*. Furthermore, it is often assumed (usually with good reason) that cistrons correspond one-to-one with specific polypeptides.

GENETIC RECOMBINATION

When a cross is performed between two organisms carrying different mutational lesions, the progeny frequently consists not only of the two parental types, but also of other types which represent new gene combinations. Genetic recombination is then said to have occurred. Recombination occurs by the breakage and rejoining of homologous chromosomes. A limited amount of DNA synthesis probably occurs in the immediate neighborhood of the rejoined ends, but elsewhere the recombinant chromosomes consist of unmodified segments of parental chromosomes. The break-reunion mechanism of recombination has been demonstrated in bacteriophage systems by means of density labeling. A typical experiment using bacteriophage λ is described in Figure 3-3 (Meselson, 1964).

Few details are known about the mechanism by which recombination occurs. In higher organisms, and at least sometimes in bacteria, the exchange is reciprocal: two parental chromosomes generate two recombinant chromosomes. This appears not to be the case in bacteriophages: among the virus particles released from single cells, the presence of one recombinant type is not correlated with the presence of the other recombinant type (Hershey and Rotman, 1949; Bresch, 1955). In the bursts from large numbers of complexes, however, these inequalities average out, and reciprocal recombinant types appear with equal frequency. The accuracy with which the recombinant chromosome is assembled very strongly suggests that chromosome homologies are tested sometime during the recombination process, presumably by what is termed homologous pairing. However, only a tiny portion of a chromosome may be actively engaged in

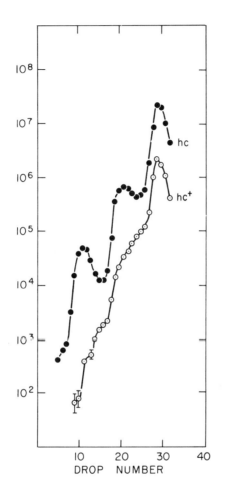

FIGURE 3-3. *Density gradient equilibrium distribution of the progeny virus particles from a cross (h$^+$c$^+$ × hc) between two density-labeled strains of bacteriophage λ (Meselson, 1964). The parental particles were grown in a medium whose nutritive components contained C^{13} and N^{15}, whereas the cross was performed in a "light" medium (C^{12} and N^{14}). The nonrecombinant (hc) particles are distributed in three well-defined regions, corresponding to heavy (unreplicated), hybrid, and "light" progeny particles. The recombinant (hc$^+$) particles are found at all regions; those which are essentially completely density-labeled demonstrate the process of recombination by break-reunion. Both distributions are broadened from three perfect modes because of recombination events — either within or away from the genetically marked region.*

homologous pairing at any time (Drake, 1967). Homology, of course, implies the existence of (nearly) identical base pair sequences. The following hypothetical scheme (Figure 3-4) represents a reasonable summary of present data concerning the molecular mechanism of recombination; more extended reviews of the problem are presented by Meselson (1967) and by Thomas (1966*b*).

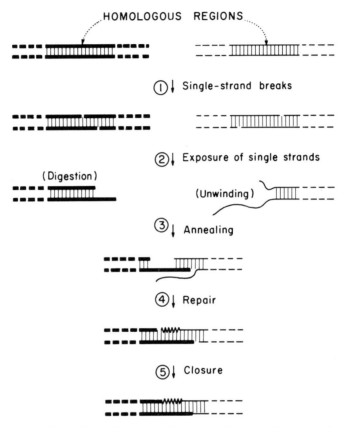

FIGURE 3-4.   *Hypothetical scheme for recombination by a break-reunion mechanism. Newly synthesized DNA is indicated by the zigzag line segment.*

(1) Single-strand breaks occur in DNA molecules. Eventually, four such interruptions must occur (even when recombination is not reciprocal), but their order and possible coordination is unknown.

(2) Single-stranded regions are generated in the regions of the breaks. They could be produced by nuclease degradation or by unwinding, but unwinding would probably also have to be a catalyzed process (Weiss and Richardson, 1966). The minimum exposed single-strand length for an organism such as T4 would have to be about 20 bases in order to ensure specificity (Thomas, 1966), but much longer regions might in practice be exposed.

(3) Annealing occurs between complementary single-stranded regions from different molecules. This can be a rapid and efficient process at the DNA concentrations within the cell. The result is a hybrid molecule, and in particular a hybrid region within that molecule. The recombinant molecule is now joined only by hydrogen bonds. The hybrid molecule may be imperfect in two ways: it may contain both single-stranded and three-stranded regions.

(4) Repair processes restore the hybrid molecule to normal complete base pairing. Three-stranded regions must be trimmed down to two strands, and single-stranded regions must be filled in, presumably by DNA polymerase.

(5) The hybrid molecule is closed by covalent bonds. The final closure requires special enzymes (Richardson, 1965; Weiss and Richardson, 1967).

The above model describes a possible mechanism of recombination for double-stranded DNA. It is commonly observed that viruses containing single-stranded DNA recombine at markedly lower rates than do viruses containing double-stranded DNA (Tessman and Tessman, 1959). Probably only the double-stranded intermediates of single-stranded viruses engage in recombination; furthermore, special restrictions seem to be placed on the double-stranded intermediates, either impeding recombination, or else impeding the replication or maturation of recombinants.

BACTERIOPHAGE CROSSES

A bacteriophage cross is performed by infecting the same cell with different bacteriophage mutants, and measuring the frequencies of recombinant viruses in the yield. Phages lack sexuality: *any* two (or more) particles may recombine within a cell. In order to perform reasonably reproducible crosses, the following procedure is widely employed, with minor variations from laboratory to laboratory.

(1) The host cells are infected with a sufficiently large average number of viruses per cell to ensure that most bacteria receive several of each parental type. The **multiplicities of infection** usually range from four to eight, and except in special circumstances, equal numbers of the two parental types are used. The particles are adsorbed as synchronously as possible, as described previously, since many bacteriophages induce resistance to superinfecting particles soon after the first particle infects a cell.

(2) The bacterium-bacteriophage **complexes** are diluted and incubated in an appropriate nutritional environment. When necessary, those parental phages which failed to adsorb can be eliminated by specific antisera or by washing the complexes.

(3) Soon after the end of the normal rise period, lysis is completed by adding chloroform. If the culture is not treated at this time, a fraction of the cells may

continue to synthesize virus; recombinant frequencies in such *lysis-inhibited* cultures are abnormally high.

(4) The frequencies of the parental and recombinant virus types are determined by appropriate assays, and the cross is considered to be acceptable if the burst size and output gene ratios fall within certain limits.

A number of aspects of bacteriophage crosses are unique to these organisms. If a cell is mixedly infected with three different mutants, the yield may contain recombinants carrying all three markers; this is a **tri-parental** cross. At times as many as ten or a dozen different parental viruses may be used in the same cross, designated an orgy cross by S. Brenner. If the complexes are prematurely broken open shortly after the end of the eclipse period, the first-matured particles are observed to contain a reduced frequency of recombinants. These and other observations clearly indicate that multiple rounds of recombination occur within infected cells. The algebraic description of this process was first laid out by Visconti and Delbruck in 1953, and was generalized by Steinberg and Stahl in 1958; it will be described briefly in the next chapter.

Genetic crosses, both in bacteriophages and in many other organisms, frequently reveal a tendency toward a nonrandom grouping of recombination events. For historical reasons this phenomenon is known by the grammatically farcical expression **negative interference**. Nonrandom clustering of recombination events in bacteriophage crosses occurs in two distinct ways. In the first case, all factors which tend to produce inequalities in the recombination potential in individual complexes will cause the yield from the "sexier" complexes to be more highly recombinant, and this distribution will produce an apparent clustering of recombination events. (For instance, two phages infecting opposite ends of a long cell might recombine more poorly than two phages infecting side by side.) In the second case, a recombination event occurring at any point on a chromosome greatly increases the chance of another recombination occurring nearby. The first case is called ordinary negative interference, and is peculiar to bacteriophage crosses. The second case is called high negative interference, and is observed in most organisms whenever very close markers are studied.

Peculiarities of particular mutants may also perturb recombination frequencies. When these marker effects are due to selective disadvantages, they are usually easily recognized. When they are due to direct effects upon recombination rates, however, they may be much less easily detected (Bernstein, 1967). A number of marker effects which depend upon specific local base pair configurations, such as AGA *versus* GAA, have been observed during fine-scale mapping of the *E. coli* tryptophan synthetase *A* gene (Yanofsky et al., 1964).

HETEROZYGOSIS

The hybrid molecule which is produced by a recombination event (Figure 3-4) contains a region in which complementary strands are derived from different parental molecules. If one of the markers in a cross happens to fall within such a region, a heteroduplex heterozygote is produced. When the heterozygote replicates, two homozygous progeny chromosomes are produced. If the heterozygote is matured before replicating, however, the particle can later initiate an infection which produces a burst comprised of two different types of particles. It is frequently easy to detect heterozygotes, depending upon the nature of the marker. If one bacteriophage enters a cross carrying an r marker and another enters carrying the r^+ marker, then the heterozygotic progeny, designated r/r^+, produce distinctively mottled plaques (Figure 3-5).

A DNA molecule clearly cannot possess normal base pairing at the heterozygous site. In fact, the molecular nature of a mutational lesion is sometimes initially revealed by effects of the lesion upon heterozygote formation.

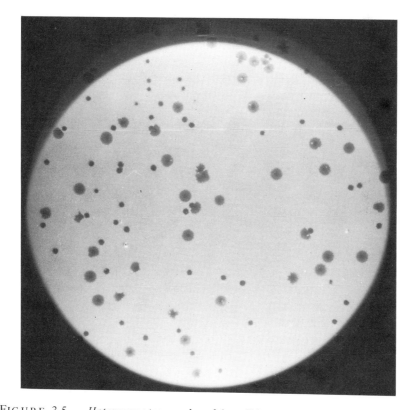

FIGURE 3-5.   *Heterozygotes produced in a T4 cross. The large plaques were initiated by r particles, the small plaques by r^+ particles, and the irregular mottled plaques by r/r^+ heterozygotes. The cross (r × r^+) was performed in the presence of FUDR, which inhibits DNA replication and therefore favors the accumulation of recombinational heterozygotes. (Cross performed by D. Lindstrom; photo by J. Sprague.)*

Since heteroduplex heterozygotes are formed by recombination and resolved by replication, their frequency soon reaches an equilibrium value. The equilibrium value for T4 is reached before the end of the eclipse period, and is about 1 percent for markers which consist of simple DNA base pair substitutions. However, agents which suppress DNA replication but not recombination can drastically increase hetrozygote frequences (Séchauld et al., 1965).

Heteroduplex heterozygotes are produced in a variety of organisms by recombination, and also, as we shall see, by mutation. However, a second type of heterozygote is also produced in bacteriophage T4, but not in higher organisms, and in fact not even in most other bacteriophages. T4 is distinguished by the fact that its terminal redundancy may encompass any portion of its genome (Chapter 2). Recombination and terminal redundancy may therefore produce a bacteriophage carrying, for instance, an *r* marker at one end and an *r*⁺ marker at the other. Such **terminal redundancy heterozygotes** differ from heteroduplex heterozygotes in a number of ways. First, they are not resolved by replication, but rather by recombination, as the reader may infer from Figure 2-4. Second, they do not produce violations of strand complementarity. Third, their formation is relatively indifferent to the molecular nature of the marker, except when rather large deletions are involved. The equilibrium frequency of terminal redundancy heterozygotes in T4 is about 0.5 percent. In contrast to hetero-

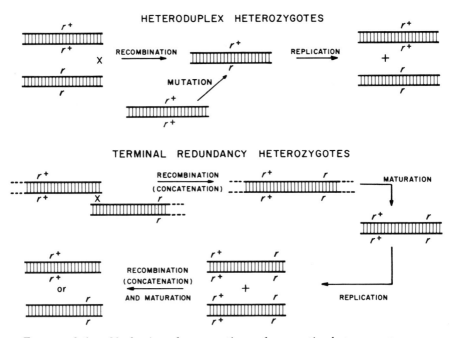

FIGURE 3-6.   *Mechanisms for generating and segregating heterozygotes.*

duplex heterozygotes, this value is not altered by suppressing DNA replication (Séchaud et al., 1965; Shalitin and Stahl, 1965).

Properties of these two types of heterozygotes are summarized in Figure 3-6. In addition, T4 sports a much smaller frequency of heterozygotes which probably result from disorders of maturation. Unlike the two types already described, the additional heterozygotes may be fractionated from the bulk of the particles, either by density gradient equilibrium centrifugation (Doermann and Boehner, 1964), or else by zonal centrifugation (Séchaud et al., 1965).

PHENOTYPIC MIXING

The **phenotype** of an organism is the sum of its immediately observable properties as expressed without further replication or genetic recombination. The **genotype** of an organism refers to the nature of its genes, and may require replication or even genetic recombination to be fully revealed.

When a bacteriophage cross is performed using a host range marker, the progeny viruses are scored according to their abilities to infect ordinary and resistant host cells. Resistant cells are generally unable to adsorb wild type virus particles, but do adsorb host range mutant particles. A considerable proportion of the progeny of such a cross is distinguished by being able to infect a resistant cell, yet by producing only wild type particles within that cell. Conversely, other particles from the cross are unable to infect a resistant cell, yet produce particles with extended host range when they infect an ordinary cell. The explanation for this result is to be found in the process of viral maturation. Both wild type and mutant tail fibers are produced in the mixedly infected cell, but they are attached to bacteriophages at random, and not according to the type of DNA in the head. In general, any character which is expressed as viral somatic protein seems to be subject to this process of **phenotypic mixing**. Genes which only express themselves intracellularly, however (such as the *r* genes), do not exhibit phenotypic mixing.

Phenotypic mixing can easily complicate the scoring of mutations. If *h* mutations are induced in the DNA of wild type particles, they cannot be scored directly by plating on resistant host cells, since the particles are still encapsulated in wild type protein. In both recombinational and mutational studies, however, phenotypic mixing can be overcome by passaging the viruses once at low average multiplicities of infection (0.1 or less), after which the genotype and the phenotype will generally correspond (except in the fairly rare case of heterozygosis).

4. Genetic mapping and the dissection of the gene

The analysis of mutational mechanisms in bacteriophages involves both the induction and detection of mutations, and also the localization of mutational lesions on the viral chromosome. Localization is measured by recombination tests with other mutants. Recombination frequencies are characteristically constant for any particular pair of mutants, but may vary over many orders of magnitude with different pairs of mutants.

ORDERING MARKERS

The chromosomes of bacteriophages, and indeed of all organisms, are simple linear structures (except that their ends are sometimes joined together to make them circular, and during replication they contain Y-shaped replication points). The first task in mapping, then, is to determine the order of various markers along the chromosome.

Consider the behavior of three markers, a, b, and c, when crossed pairwise in all three possible combinations. (For simplicity, we will abbreviate the description of these **two-factor crosses** by condensing $ab^+c^+ \times a^+bc^+$ to $a \times b$, and so on.) Each cross will produce a characteristic frequency of recombinants (the sum of the frequencies of the two reciprocal recombinant types). These frequencies, which we will call $R(ab)$, $R(bc)$, and $R(ac)$, are in fact the probabilities that odd numbers of exchanges occur within the specified intervals. An even number of exchanges between two markers will, of course, fail to recombine them. If one assumes initially that genetic exchanges are distributed at random along the chromosome, then the magnitude of R will depend upon the distance between markers. The markers can then be ordered, since the largest R value should correspond to the two outermost markers. In addition, the markers can be approximately scaled by placing the middle marker at a point which divides the distance between the outside markers in proportion to the two smallest R values. The result is a primitive **genetic map** (Figure 4-1).

An entirely different method of mapping is possible using deletion mutations and qualitative rather than quantitative recombination tests. **Deletion mapping**

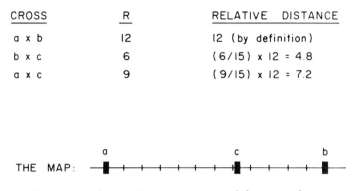

CROSS	R	RELATIVE DISTANCE
a x b	12	12 (by definition)
b x c	6	(6/15) x 12 = 4.8
a x c	9	(9/15) x 12 = 7.2

THE MAP:

FIGURE 4-1.   *A simple genetic map constructed from two-factor crosses. R =*
percent recombinants

will be described later in connection with fine-scale mapping, but it is applicable
in principle over either long or short distances.

R-value mapping tends to fail with markers which are so far apart that, on the
average, several exchanges occur between them. In this case all R values
approach 0.5 (the probability of an odd number of exchanges). There still
remain other ways to order such distant markers, however. With luck, the
interval between them may be subdivided by other markers. Alternatively, the
average number of exchanges at all points can be decreased by premature lysis of
the complexes, thus shrinking the entire genetic map. In general, marker pairs
which exhibit R values less than 0.5 are said to be linked. Marker pairs which
exhibit R values close to 0.5 may or may not be linked. Two-factor crosses may
also produce a map which is not colinear with the chromosome over very short
distances, such as those encountered in fine-scale mapping (D. Zipser, personal
communication).

Consider now a set of three two-factor crosses which produce an anomalous
result, such as three identical R values, or a map order inconsistent with other
previously mapped markers. The markers may still be ordered in a simple,
one-dimensional map by means of a **three-factor cross**. The cross $ab \times c$, or any
of the three other possible three-factor crosses, will produce eight genotypes
(two parental genotypes plus six recombinant genotypes). These can be grouped
into four classes, each containing reciprocal genotypes (such as abc and $a^+b^+c^+$).
Two of the three recombinant classes result from a single exchange, while the
third results from a double exchange (Figure 4-2). The frequency of the
double-exchange class is virtually always smaller than the frequency of either
single-exchange class, and the marker which required two exchanges to be moved
between chromosomes must be the central of the three markers.

Two difficulties arise in three-factor crosses in microbial systems. First, a
double mutant must be constructed; in microbial systems, R values may be very
small, and enrichment selection of the double mutant may not be feasible.
Second, the eight genotypes may not present enough easily scored phenotypes

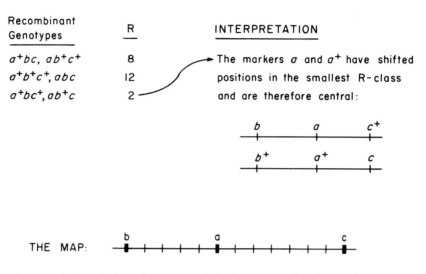

ANALYSIS OF THE THREE-FACTOR CROSS $abc^+ \times a^+b^+c$

Recombinant Genotypes	R	INTERPRETATION
a^+bc, ab^+c^+	8	The markers a and a^+ have shifted
$a^+b^+c^+, abc$	12	positions in the smallest R-class
a^+bc^+, ab^+c	2	and are therefore central:

FIGURE 4-2.   *A three-factor cross. The frequency of exchanges between a and b is $(8 + 2) = 10\%$; between a and c it is $(12 + 2) = 14\%$.*

to determine the frequencies of the three recumbinant classes. For instance, if all three markers are located within the same gene, then only two phenotypes may occur: mutant and wild type. In practice, fortunately, two-factor crosses usually suffice to order the markers used in bacteriophage crosses.

GENETIC MAPS

A genetic map is constructed not only to indicate the order of markers, and hence presumably their arrangement along the chromosome, but also to indicate quantitatively their distances from each other. Moderately accurate maps can often be put together simply from the results of a number of two-factor crosses. On the other hand, numerous corrections can be made of crude R values to account for the effects of various biological or statistical factors which may distort recombination frequencies. Quite detailed maps are now available for bacteriophage T4, which has been more closely mapped than has any other organism. The T4 map is circular (Figure 4-3), a result of circularly permuting linear chromosomes. Maps of this type can be constructed by choosing some particular recombination frequency as unity, and then either constructing the map from small multiples of this frequency, or else by applying appropriate correction factors. The conventional map unit for most bacteriophages corresponds to a recombination frequency of 1 percent.

A detailed understanding of recombination should make possible the construction of an algebraic relationship between R values and a linear variable d

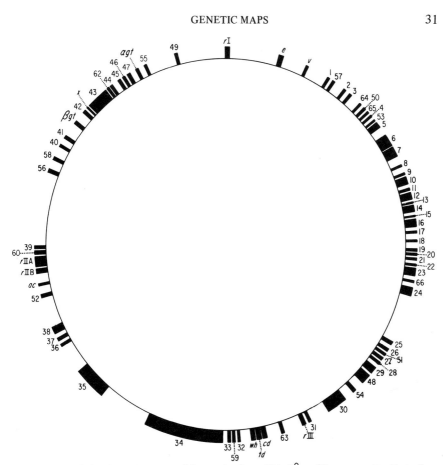

FIGURE 4-3. *Genetic map of bacteriophage T4: 3° ≈ 20 map units. Data from Edgar (personal communication), Georgopoulous (1968), Hall and Tessman (1967), Stahl, Edgar, and Steinberg (1964), and Wood and Edgar (1966).*

which is proportional to true chromosomal distances. An equation of this sort is called a **mapping function**. A fairly simple mapping function, first constructed by Visconti and Delbrück in 1953, describes R values in T-even phages for values less than 0.5. It can be written in the form

$$R = \frac{1}{2}(1 - e^{-md}) \tag{4-1}$$

where m is the average number of mating events in the lineage of any progeny particle. A more detailed mapping function was developed by Hershey in 1958:

$$R = 2abf_1 f_2 \left(1 + \frac{e^{-m_2 d} - e^{-m_1 d}}{d(m_2 - m_1)}\right). \tag{4-2}$$

Here a and b are the relative multiplicities of the parental viruses $(a + b = 1)$, f_1 is a correction factor for the effects of the random distribution of infecting particles among different cells, f_2 is a correction factor for the effects of nonrandom distributions of exchanges in lineages, and m_1 and m_2 are the average numbers of matings for the first-matured and the last-matured particles, respectively.

Probably the most important aspect of mapping functions for the student of mutation is the possibility they offer of correlating R values with physical distances, that is, with numbers of base pairs between (or within) markers. An extensive study by Stahl, Edgar and Steinberg (1964) resulted in such a relationship, which is reproduced in Figure 4-4. Notice that the T4 map unit ($R = 1$ percent) corresponds to about 104 base pairs. This number agrees well with two other observations. First, the outermost markers of the T4e gene produce about 3 percent recombinants when crossed, and the e gene encodes a lysozyme molecule composed of 164 amino acids (Tsugita and Inouye, 1968), which in turn correspond to 492 base pairs. Figure 4-4 predicts about 470 base pairs corresponding to $R = 3$ percent. Second, the physical distance between two markers in T4 has been estimated (Goldberg, 1966). Bacteriophage DNA was extracted from virus particles, melted to the single-stranded state, treated with endonuclease, and then assayed in a DNA transformation system. The nuclease treatment produced known numbers of randomly located breaks in the DNA

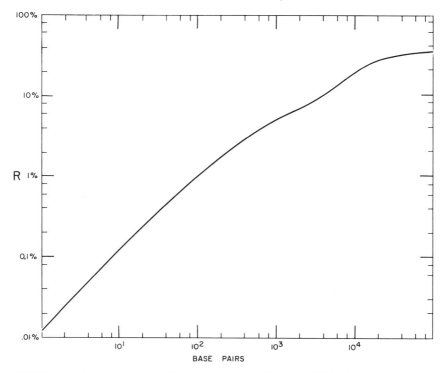

FIGURE 4-4.   *A mapping function for bacteriophage T4 (redrawn from Stahl, Edgar and Steinberg, 1964).*

molecules, so that the probability of separating two linked markers could be transformed into a physical distance. The measured interval contained about 2700 base pairs and about 19 map units, or 140 base pairs per map unit.

The preceding discussion has tacitly assumed an invariant probability of recombination per base pair, but this is not always the case. The physical distances corresponding to genes 34 and 35 (Figure 4-4), for instance, are nowhere as large as their map lengths would indicate (Mosig, 1966), implying the presence in these genes of exceedingly recombination-rich regions. Other areas of the map appear to be subject to similar but smaller variations (Rottlander et al., 1967; Berger, 1965, but see also Drake, 1966a).

FINE-SCALE MAPPING

In mutational studies, attention is frequently focused upon markers which map very close to each other. Mapping many markers within a gene, particularly when performed with near-nucleotide resolution, is called fine-scale mapping. In addition, the term has often come to imply the mapping of large numbers of mutants within only one or a few genes.

Extensive mapping of closely linked markers requires selective methods for recording recombinants. The viral markers most often employed are conditional lethal mutations. These have the ability to grow well under one set of conditions, but not under another set of conditions which do permit the growth of the wild type. (The differential growth conditions are called **permissive** and **nonpermissive**, respectively. Examples are: high and low temperatures; sensitive and resistant host cells; and drug and radiation resistance.) Usually only wild type recombinants are scored under nonpermissive conditions; their frequency is doubled to obtain an R value. This procedure may be faulty in two ways, however. Heterozygotes, whether of the heteroduplex or of the terminal redundancy type, will be scored as recombinants with a probability which depends both upon the particular gene, and upon the molecular nature of the mutational lesion within it. For instance, a heteroduplex heterozygote for a marker in an "early" gene will tend to be scored only if the DNA strand which is *not* used to determine the base sequence of the messenger RNA is the mutant strand. In the "late" gene this problem does not arise, since the heterozygote will by then have been replicated (Hertel, 1965). On the other hand, a marker which consists of a deletion of a few base pairs may disrupt the reading even of the wild type strand in a heteroduplex heterozygote, since this strand will be looped out opposite the mutational lesion. Unfortunately, it is generally not possible to predict such transmission probabilities with confidence. One way out of this difficulty, of course, is to passage the lysate once at a low multiplicity before determining recombination frequencies: most heterozygotes will then be resolved.

It is also conceivable, and indeed occasionally likely, that certain pairs of very close markers produce reciprocal recombinant types at different rates. At present it is impossible to predict when this will occur, and in fact it is usually difficult for technical reasons to measure nonreciprocality. Recombination values which are based only upon wild type recombination frequencies must therefore be accepted with reservations.

With the assistance of mutagens and selective screening procedures, several hundred viral mutants may easily be obtained within a few days. Consider now the mapping of a hundred newly isolated mutants by means of pairwise crosses. The number of possible crosses is rather large: $99 + 98 + 97 + \ldots = 4950$. An industrious and unimaginative person might reasonably perform and analyze 20 crosses per day, and would therefore consume a normal working year mapping the one hundred mutants. Together with materials, this map would run to about $12,000, plus indirect costs.

There is a better way. It depends upon the properties of deletion mutants which have lost considerable numbers of contiguous nucleotides. Deletion mutants are recognized by two criteria: they fail to revert to the wild type, and they fail to recombine with two or more other mutants, which can themselves recombine to produce wild type. When a number of deletions become available, these may be crossed with a few point mutants which more or less span the region whose fine-scale mapping is contemplated; the ends of the deletions may thus be roughly determined. With some luck, a set of deletions can be found whose ends subdivide this region extensively. Figure 4-5 shows a set of deletions which subdivides the two rII genes of bacteriophage T4 into a large number of segments (Benzer, 1961a). Similarly useful sets of deletions have been employed in $E.\ coli$ for the mapping of the β-galactosidase z gene (Beckwith, 1964; Jacob, Ullmann, and Monod, 1964) and the tryptophan synthetase A gene (Yanofsky et al., 1964).

This set of deletions now becomes a powerful tool for further mapping. Each point mutant is first crossed against several of the deletions. A standard cross is not used, however, because it is only necessary to determine whether *any* recombination occurs, not how much. The simplified cross which is used is called a recombination spot test, and is described in the legend to Figure 4-6. (Scores of recombination spot tests can be performed in the time required to perform a single standard cross.) Any mutant which recombines with a deletion is clearly located outside of the deleted region. The pattern of recombination with a set of deletions thus localizes a point mutant within a particular segment.

Depending upon the number and location of the mapping deletions, the various segments will receive various fractions of the collection of point mutants. (If any segment receives a large number of mutants, further deletions may be sought to subdivide that segment.) The next step involves making pairwise crosses of all of the mutants falling within a segment, *again using recombination spot tests*. The mutants within a segment are thus grouped into **sites**. (Two mutants are said to occupy the same site when they fail to recombine with each other to produce wild type recombinants.) Finally, if required, representatives of each site within a segment may be crossed with each other in standard

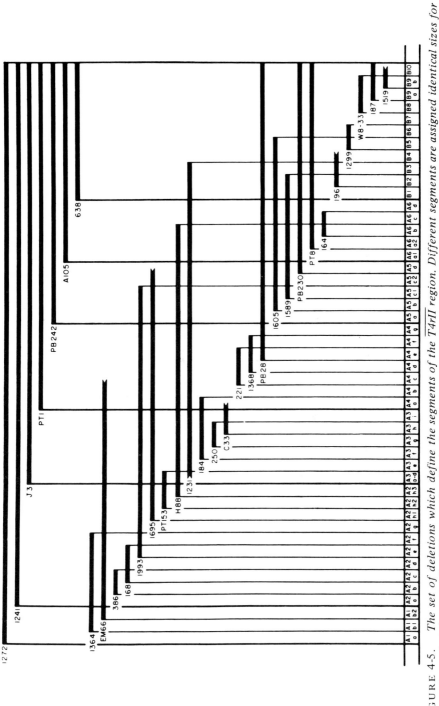

FIGURE 4-5. *The set of deletions which define the segments of the* T4rII *region. Different segments are assigned identical sizes for convenience, but they actually differ considerably in size.*

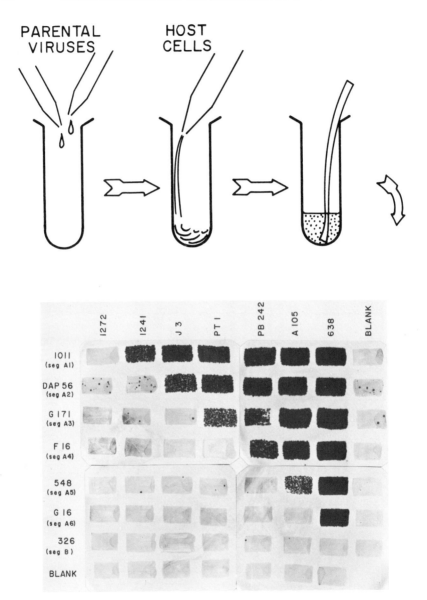

FIGURE 4-6. *Recombination spot tests. A drop of each parental T4rII mutant at about* 4×10^9 */ml is added to a small tube, followed by about 0.5 ml of E. coli B cells at about* 10^8/ml. *After about 10 minutes at room temperature, the mixtures are spotted with sterile paper strips onto a prepoured lawn of K(λ) cells. After overnight incubation, clearing (dark areas) results from recombinants produced in the B cells, while occasional, isolated plaques usually result from revertants. The mutants listed across the top of the grid are mapping deletions, while those listed down the left side are point mutants.*

two-factor crosses. The total number of standard crosses required has now been drastically reduced. If a hundred mutants divide equally among ten segments, and if each segment contains five sites, then a hundred two-factor crosses would be required. This is 2 percent of the number required without recombination spot tests and deletion mapping.

Two limitations exist to the resolving power of fine-scale mapping: the efficiency of screening of recombinants, and the stability of point mutants. Screening procedures for wild type particles do not in general distinguish between those which arose from recombination, and those which arose from back-mutation. A pair of markers cannot, therefore, be mapped more closely than the distance which is determined by their reversion rates. If this "distance" is well below the recombination frequency expected between truly adjacent markers, it will not interfere with fine-scale mapping. As will be seen below, however, the minimum recombination frequency is more difficult to predict that Figure 4-4 would imply.

EXTREMELY CLOSE MARKERS

The T4rII region has been more closely mapped than has any other region in any organism. This monumental task was performed by Seymour Benzer and his associates, and will be described in more detail in later chapters. These workers attempted very early in their analysis to estimate minimum recombination frequencies. Making the very reasonable assumption that recombination never subdivides a base pair, and confining our attention for the time being to markers which consist of base pair substitutions and not of base pair additions or deletions, then the elementary physical units corresponding to genetic distances should of course be the internucleotide gaps, and not the base pairs themselves. For the first several years, the smallest R value encountered in the T4rII region was about 2×10^{-4} (Benzer, 1957). From Figure 4-4, the corresponding nucleotide distance between the markers would be about 2; but if the "base pairs" scale is interpreted to mean internucleotide gaps, then a single base pair would lie between the two markers.

Tessman (1965) later observed that mutants which had originally been assigned to the same T4rII site on the basis of recombination spot tests, could often be differentiated phenotypically. Different members of a site often exhibited different abilities to grow on certain partially permissive host cells. Very carefully conducted standard crosses then revealed that some of these mutants could recombine at extremely low frequencies: r^+ frequencies as low as 10^{-8} were observed. In terms of Figure 4-4, the corresponding R values would imply not only the subdivision of DNA base pairs, but also of their very atoms. It is conceivable that these apparently wild type particles are produced by some special mechanism which differs from ordinary recombination (for instance, by a special process of mutation), but until the situation is more perfectly understood, the criterion for assigning mutants to particular sites should be made very stringent: the mutants should fail to recombine *in the most sensitive tests available.* (Note, however, that phenotypically distinct mutants *can* map at the same site, which may be occupied by as many as four different base pairs.)

Recombination has also been measured between markers which were already known to consist of base pair substitutions at adjoining base pairs. Crosses were performed between mutant strains of *E. coli* carrying two different amino acid substitutions in their tryptophan synthetase A proteins, for instance, between a *cys* strain and an *asp* strain, both derived from the wild type *gly* strain; here the cross must have been ... UGP ... X ... GAP ..., where P is a pyrimidine. Recombination was observed at only about 20 percent of the expected rate estimated from crosses between quite distant markers; by contrast, recombination between markers separated by three to six internucleotide gaps occurred at about 150 percent of the expected rate (Yanofsky et al., 1967). Recombination within the T4rII region within homologous UGA and UAG codons occurred with an R value of $(0.4-1.2) \times 10^{-4}$ (Brenner et al., 1967), virtually as predicted by Figure 4-4. When the same UGA X UAG cross was performed using the β-galactosidase gene of *Escherichia coli*, recombination was about 10-fold lower than expected (Zipser, 1967). It is therefore clear that adjacent markers may recombine at anomalously low rates, but that they can sometimes recombine at or near the expected rate.

5. Mutation rates

Accurate measurements of mutation rates are frequently required for resolving mutational mechanisms and for characterizing particular mutational sites. The measurement of a mutation rate requires not only an efficient method for detecting and counting mutants, but also an understanding of the factors upon which the mutation rate depends. Here we will consider mutation rates as time-dependent or generation-dependent functions. The general principles which apply in collecting independently arising mutant individuals will also be reviewed.

FORWARD AND REVERSE MUTATION

It is appropriate at this point to describe the difference between forward and reverse mutation. It is usually tacitly assumed that a wild type organism possesses all of its genetic functions intact. This is especially true for haploid organisms, which generally possess only single copies of each of their genes, and are not in a position to carry concealed mutational damages. Under these conditions, the heritable alteration or loss of some gene function is attributed to a **forward mutation**, which may occur at any one of a large number of sites within a gene or a gene cluster.

When a mutant individual is grown into a large population, secondary mutations which restore whatever activity was lost at the time of the original mutation may often be observed. Such events are called **reversions** or **back mutations**. They may occur either at the originally mutated site, or else at some other site. If reversion occurs at the original site and restores the original wild type DNA configuration, it is a **true reversion**. If reversion occurs either at some secondary site, or at the original site but not so as to restore the original wild type DNA configuration, it is a **false reversion**, or pseudoreversion, or partial reversion. When the reversion definitely occurs at a site separable from the originally mutated site by recombination, it is said to occur by **suppression**, or by a suppressor mutation.

The main difficulty with these formulations of forward and reverse mutation arises from the concept of a wild type gene. One might naively suppose the unmutated gene to consist of some optimally effective DNA base pair sequence.

In fact, however, a considerable number of nearly optimally effective versions of a gene may exist among wild type individuals, and differing environmental conditions may alternatively favor one or another of these genes. This situation can sometimes lead to very unexpected results in mutational studies. Consider, for example, a mutant strain of *E. coli* selected for resistance to bacteriophage T2, the resistant cell strain being used to select extended host range mutants of T2. (The resistant cells derived from *E. coli* strain B would be called B/2. The host range mutants of T2 would be called *h1, h2, h3,* and so on in order of isolation.) When crossed among themselves, it turns out that different *h* mutants do not recombine to produce h^+ wild type particles. On the other hand, a given *h* mutant reverts to h^+. Independently arising h^+ revertants turn out to be capable of recombination, one of the products of this recombination being the original *h* "mutant". In this case forward mutation clearly occurs at a single site only, whereas reversion occurs at several sites. Of course, many other "forward" mutations probably do occur, but are only detected if they are conditional lethal mutations.

MUTATION RATES IN NONREPLICATING SYSTEMS

It is impossible to separate a discussion of mutation rates from a discussion of mutational mechanisms. Consider, for example, the effects of nitrous acid upon a virus. Various types of mutations may be produced, and at the same time large numbers of the particles will be inactivated. What are the relevant variables which determine the mutation rate? Since free virus particles are treated, and since the mutations which are induced among the survivors appear to be expressed more or less immediately upon infection, numbers of viral generations are clearly irrelevant. Time of treatment would appear to be a more reasonable independent variable. Under certain conditions, however, nitrous acid may be sufficiently unstable to maintain a steady concentration during treatment, and some other variable must be employed. On the other hand, when viruses are treated with the alkylating agent ethyl methanesulfonate, the numbers of mutations produced depends not only upon the length of treatment, but also upon the number of subsequent viral generations occurring before the mutants are counted. Clearly some a priori picture of the details of the mutation process are required in order to make a choice of the most suitable method of expressing mutation rates.

Under conditions where mutation depends linearly upon time, such as in the presence of a large excess of a stable chemical mutagen, the mutation rate will be described by

$$M - M_0 = \mu t \tag{5-1}$$

where M is the time-dependent mutant frequency, M_0 is the zero-time mutant frequency, μ is the mutation rate, and t is time. This expression will only be valid, of course, when M is small.

When the mutation rate varies with time, or when it is constant but it is desirable to compare experiments performed under slightly different conditions or in different laboratories, a more suitable expression can be derived by assuming that mutational events and lethal events are independent but proportional. This assumption appears to be satisfactory in most instances. The distribution of lethal events (called **lethal hits**) among the members of a virus population is usually well described by the **Poisson distribution**,

$$P(x) = \frac{h^x e^{-h}}{x!},$$

(5-2)

where h is the average number of lethal hits per virus particle, e is the base of the natural logarithms (2.71828 . . .), and $P(x)$ is the probability that a virus receives exactly x lethal hits in some sensitive target. (Remember that $0! = 1$. Note also that a virus particle may receive several lethal hits.) In the case where a single hit is sufficient to inactivate the ability of a virus particle to initiate infection, the proportion of surviving particles (S) is given by

$$S = P(0) = e^{-h}.$$

(5-3)

In the case where inactivation depends upon a hit within each of the n targets within the particle, the proportion of surviving particles is given by

$$S = 1 - (1 - e^{-h})^n.$$

(5-4)

(This is a **multiple-hit** equation, derived as follows: the probability of a single target surviving is e^{-h}. The probability of its inactivation is therefore $1 - e^{-h}$. The probability of inactivating all n targets in a given particle is $(1 - e^{-h})^n$. Any other outcome leads to survival. If the hits are delivered at a constant rate, it is convenient to let $h = kt$ and to plot $\log_{10} S$ against time, as in Figure 5-1. The single-hit equation then becomes

$$\log S = -0.4343kt.$$

(5-5)

In the case of the multiple-hit equation, $\log S$ asymptotically approaches the straight line,

$$\log S = \log n - 0.4343kt.$$

(5-6)

The zero-intercept of this line provides the value of n. For both types of curve, the time scale can easily be converted to a hit scale, as illustrated in Figure 5-1. For the single-hit curve, $h = 1$ corresponds to $S = 0.37$, and $S = 0.1$ corresponds to $h = 2.303$. Similar considerations apply to the multiple-hit curve, except that the value of n must be known.

Analyses of lethal and mutational processes frequently reveal multiple-hit kinetics, but it is important to note that the underlying mechanisms may be less obvious than might be suggested by the derivation of equation (5-4). Very often, for instance, a considerable proportion of lesions are repaired at low doses, but with increasing doses, the repair system becomes either saturated or inactivated.

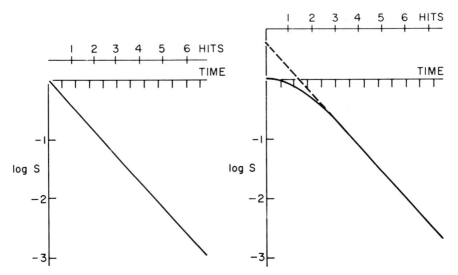

FIGURE 5-1.   *Inactivation curves. Left: single-hit kinetics. Right: multiple-hit (about 4-hit) kinetics.*

The saturation of a repair system may be detected by a strong dose rate effect: a given dose has a much greater effect when applied in a very brief time.

It is usually a simple matter to determine the surviving fraction of a treated virus population (and its inactivation kinetics, if not already known). It is also usually simple to determine the frequency of mutants per survivor. Thus the mutation rate can be expressed in terms of mutational hits per lethal hit:

$$M - M_0 = Kh = -2.303\ K \log S \tag{5-7}$$

where K is the ratio of mutational to lethal hits, and where the final part of the expression holds only for single-hit inactivation kinetics. As an example, bacteriophage T4 is inactivated by nitrous acid with single-hit kinetics, and at the same time the reversion $rN24 \to r^+$ is induced. A plot of mutant frequency against log of surviving fraction of viruses is linear (Figure 5-2); for examples see Granoff (1961) and Bautz-Freese and Freese (1961). On the other hand, T4 is inactivated by ethyl ethanesulfonate with multiple-hit kinetics. In this case, a plot of mutant frequency against log of surviving fraction of viruses bends distinctly downwards (Bautz and Freese, 1960; Freese, 1961), since the expression $M - M_0 = -2.303\ K \log S$ should be replaced by the more cumbersome expression, $M - M_0 = -2.303\ K \log\left(1 - (1 - S)^{1/n}\right)$.

It becomes much more difficult to describe mutation rates when a mutagen acts by chemically modifying one of the DNA bases, and the altered base then mispairs during subsequent replications with a probability decidedly less than unity. This behavior, for instance, appears to be characteristic of alkylating agents (such as ethyl ethanesulfonate). It is convenient in practice to measure the induced mutant frequency after a precise number of DNA replications has

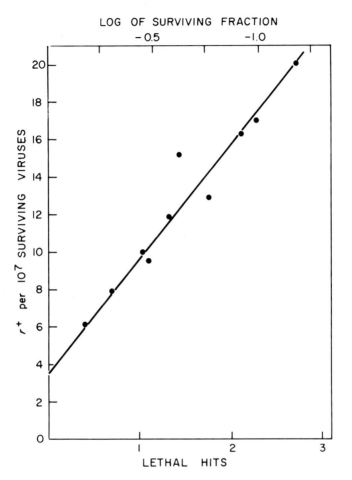

FIGURE 5-2. *Linear production of mutations (rN24→r⁺) by nitrous acid as a function of lethal hits delivered (data from Bautz-Freese and Freese, 1961).*

occurred, or conversely, to determine at which replication after treatment the mutation first appeared. In the case of the induced reversion of a conditional lethal mutation in an "early" viral gene, on the other hand, no replications will occur unless the mutation itself has already occurred.

MUTATION RATES IN REPLICATING SYSTEMS

Many mutagens specifically induce mutations during the replication of viral nucleic acids, but show no effects upon nonreplicating virus particles. Base analogues such as 5-bromouracil and 2-aminopurine are good examples. Under these conditions, mutation rates must be described in terms of mutations per generation, or per DNA replication. Lethality is generally low or nonexistent with such mutagens (except for lethal mutations themselves), and time is not an independent variable except in the guise of numbers of generations.

Unfortunately, however, there are several difficulties in the measurement of numbers of viral generations. First of all, the mode of nucleic acid replication must be understood. For viruses with semiconservatively replicating double-stranded DNA, the generation is a well defined concept. For single-stranded viruses which exhibit double-stranded intermediates, however, the "generation" is more complex: the double-stranded form may first replicate semiconservatively for a few generations, and may then proceed to produce exclusively single-stranded progeny. Furthermore, only selected members of the double-stranded intermediates may act as the templates for single-stranded progeny. In addition to these conceptual difficulties, there exist counting problems. All viral systems, for instance, exhibit a certain amount of wastage: when intracellular virus synthesis ceases (usually at the time of cellular lysis), a number of viral genomes remain unmatured; the vegetative pool is not used up at the end of the latent period. (A very dense suspension of $E.$ $coli$ which has just lysed as a result of T4 infection may momentarily turn nearly solid because of the extremely high viscosity due to released viral DNA; however, endogenous nuclease activity soon liquifies the suspension.) In calculating the factor of increase from the originally infecting particles, therefore, the number of wasted chromosomes must be added to the observed burst size.

The replication-dependent mutation process can be formulated in a number of ways. For a very simple formulation, let m = the **probability of mutation per replication**, M = the number of mutants in a population at a given time, and N = the total size of the population at that time. If it is assumed that M remains much smaller than N, that reversion does not significantly alter the magnitude of M, and that both mutant and nonmutant organisms replicate at the same rate, then the rate of increase of mutants in the population will be given by

$$dM = \left(m + \frac{M}{N} \right) dN \qquad (5\text{-}8)$$

(The increase in the mutant population is divided into new mutations, and replication of previously accumulated mutants.) The number of mutants can also be expressed in terms of the mutant frequency f by the relationship $M = fN$. Differentiation of this expression, substitution into equation (5-8), and integration yields

$$f - f_0 = m \ \ln(N/N_0) = m(\ln N - \ln N_0) \qquad (5\text{-}9)$$

where f_0 and N_0 describe the population at the beginning of the experiment.

The application of equation (5-9) may be further simplified by growing a population from a number of individuals so small that $f_0 = 0$; then a single determination of mutant and nonmutant population sizes suffices:

$$m = \frac{0.4343f}{(\log N - \log N_0)} \tag{5-10}$$

In a virus suspension obtained by inoculating 5 ml of a cell suspension with 10^2 virus particles, where the final titer (corrected whenever possible for losses) was 10^{10}/ml, the frequency of any type of mutant would thus be some 20 times larger than the mutation rate. The main difficulty with this shortcut method is its sensitivity to "jackpots." The very first mutation to occur during the growth of a culture often contributes disproportionately to the final mutant frequency. If by chance the first mutation occurs quite early in the growth of the culture, a correspondingly high final mutant frequency will result because of the continued replication of the mutant. The mutation rate is therefore best determined either by growing several cultures, and adopting the lowest value of m so obtained, or else by starting with a population so large that it is free of these fluctuations, and applying equation (5-9).

It can easily be shown, along with the derivation of equation (5-11) below, that all generations contribute equally to the final mutant frequency in a population of geometrically reproducing organisms: the exponential increase in the number of new clones initiated in each generation is compensated by the exponential growth of the previously established clones. This result can be used to estimate the likelihood of appearance of a stock with an anomalously *low* mutant frequency. The most probable course of mutant accumulation is as follows: when the population achieves size $2^0/m$, the first mutation occurs; at size $2^1/m$, 2 mutations occur; and so on, until at the final size $2^i/m$, 2^i mutations occur. The final mutant frequency

$$f = M/N = (2^0 \times 2^i + 2^1 \times 2^{i-1} + \cdots + 2^i \times 2^0)/(2^i/m) = i2^i/(2^i/m) = mi$$

which is equivalent to equation (5-10). Now, while it is possible by chance fluctuations for a population occasionally to escape mutations for small values of i, the probability of repeated escapes becomes negligible for values of i above about 5. If a population is grown to sufficiently large i-values, therefore, early random nonoccurrences of mutations become relatively less important. Thus for a population initiated by fewer than 10^5 organisms, wherein a mutation for which $m = 10^{-5}$ will be studied, growth to 10^{12} organisms will correspond to $i = 23$, and the degree of underestimation is not likely to exceed $5/23 = 17$ percent. Jackpots, on the other hand, can lead to overestimations of hundreds to thousands of percentage points.

Reliable methods for measuring mutation rates make use of the different times during the growth of a population when mutants make their appearances. In order to justify the use of these methods, however, it is first necessary to demonstrate that viruses, both mutant and wild type, replicate geometrically (Luria, 1951). A large population of viruses grown from a small, mutant-free population will contain a mixture of mutant clones, each of which arose in some previous generation. Assume for simplicity that viral replication occurs by synchronous replication; the reality of nonsynchrony, while algebraically more involved, leads to essentially the same result. Assume further that mutations which might arise as heteroduplex heterozygotes are not scored until they replicate to homozygosity. (Refer to Figure 5-3 for assistance in following the argument.) Let x be the size of a mutant clone which arose during the kth replication in the past; then $x = 2^{k-1}$. Let N be the size of the final population. A clone of size x therefore arose from a population whose size was $N/2^{k-1}$. Let m be the probability of mutation per replication, where a mutation results in one mutant and one nonmutant progeny. The number of mutant clones of size x is therefore equal to $mN/2^k$. It will be convenient, however, to calculate the number $Y(x)$ of clones which contain x *or more* mutants:

$$Y = \sum_k^\infty \frac{mN}{2^k}. \quad \text{Since} \quad \sum_k^\infty \frac{1}{2^k} = \frac{1}{2^{k-1}},$$

this expression simplifies to

$$Y = mN/x, \quad \text{or} \quad \log Y = \log(mN) - \log x. \quad (5\text{-}11)$$

A plot of log Y against log x is therefore linear, with slope of (-1) and with $Y(1) = mN$. (This expression will depend to within a factor of 2 upon whether or not mutations arise as heteroduplex heterozygotes, and whether or not such forms are actually scored as mutants.)

In practice, $Y(x)$ and N are measured as follows: cells are infected at a low multiplicity, such as 0.01, and are then diluted and distributed to a large number of individual tubes before the end of the latent period; on the average, each tube receives only a few infected cells. When lysis is complete, each tube is individually plated. Most of the plates contain numerous nonmutant plaques,

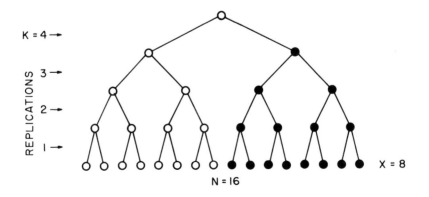

FIGURE 5-3.   *The initiation and growth of a mutant clone.*

but only a fraction of the plates contain any mutants. Plates with mutants thus usually contain only a single mutant clone. The mutant clones are individually counted, and in addition, the total number of plaques is determined on a few plates. (For studying much more rare mutational events, a variation on this procedure involves distributing a large number of infected cells to each tube, so that, on the average, a mutant clone still appears in only about 20 percent of the tubes. The mutants are then scored by some selective method, and a few tubes are also assayed for total viruses. This procedure is particularly well suited for studying the reversion of a conditional lethal mutant.) As an example, Figure 5-4 shows the results of Luria's measurements of r mutants appearing in bacteriophage T2.

Up to about $x = 48$ there is good agreement with expectation; at higher values, limitations on the burst size lead to departures from the theory. A much more realistic model of the same process was examined by Steinberg and Stahl (1961). They assumed that both replication and maturation were random-in-time processes operating upon a steady-state vegetative pool of viral nucleic acid. The resulting distribution of clone sizes could explain both the existence of clones of sizes different from 2^k and the reduced number of large clones.

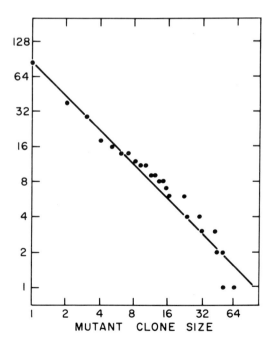

FIGURE 5-4. *The cumulative clone-size distribution of spontaneous T4r mutants (Luria, 1951). Y = number of clones containing x or more mutants, where x = mutant clone size.*

(The rather huge experiment of Luria [1951] was based on a study of 23,000 bursts, or about 1,850,000 plaques, contained on 2874 plates. A total of 85 plates contained new mutants, which corresponded to 87.6 clones when corrected for coincidences. The mutation rate $m = Y(1)/N$ was therefore about 5×10^{-5}. As a method of determining mutation rates, of course, this procedure is clearly tedious. The experiment was only feasible because of the power of the human eye to detect rare events: a plate containing over a thousand wild type plaques can be scanned in a few seconds, and both r and mottled plaques can be detected with high efficiency.)

An interesting exception to the simple model of geometrical replication has been observed in bacteriophage $\Phi X174$, where mutants appear in a random rather than a clonal distribution (Denhardt and Silver, 1966). This virus appears to employ a very limited number of templates for the production of progeny strands, and the progeny strands usually do not replicate further.

The general principles developed in the preceding paragraphs can be applied to a moderately efficient method for determining viral mutation rates. The approach, first developed by Luria and Delbrück in 1943 and later generalized by Lea and Coulson in 1949, is particularly useful when the mutations being studied produce individuals with growth rates different from that of the parental population. Furthermore, although the method scores mutant clones, it requries far fewer plates than do the measurements described previously. A number of tubes, on the order of one hundred, are seeded with host cells and with a virus inoculum sufficiently small so that preexisting mutants are rarely introduced. Viral growth is then allowed to proceed, either for a single cycle or for a number of cycles, until a point is reached when about 0.2 to 0.8 of the tubes contain at least one mutant. Simple preliminary runs are generally required to determine the necessary amount of growth. The most accurate results will be achieved when about 0.2 of the tubes contain *no* mutants (Lea and Coulson, 1949). A small number of the tubes are assayed for total viruses, the average being adopted for all of the tubes. The rest of the tubes are plated out in their entirety, one tube to a plate. These plates are then scored only for the presence or absence of mutants, regardless of the actual numbers present on a plate. (If the mutation is rare, selective measures must be used; for instance, the mutation $T2h^+ \rightarrow T2h$ would be scored on plates inoculated with B/2 cells. The method is apt not to work well for plaque morphology markers because of the difficulties of detecting mutants on very crowded plates.)

The results are interpreted as follows: let the initial and final average concentration of viruses in individual tubes be N_0 and N, respectively. Let m be the mutation rate per viral replication. Since the total number of replication acts per tube is given by $N - N_0$, the total average number of mutations M per tube will be given by $M = m(N - N_0)$. Generally $N \gg N_0$ so that

$$M = mN. \tag{5-12}$$

If mutations are truly random events, then they will be distributed among the tubes according to the Poisson distribution (5-2). In this case, the fraction of tubes with no mutants will be described by

$$P(0) = e^{-M}. \qquad (5\text{-}13)$$

Combining (5-12) and (5-13), we have

$$P(0) = e^{-mN}, \qquad m = \frac{-\log P(0)}{0.4343 \ N}. \qquad (5\text{-}14)$$

It should also be noted that (5-14) often appears in the form

$$m' = \frac{-(\ln P(0))(\ln 2)}{N} = \frac{-1.6 \log P(0)}{N}. \qquad (5\text{-}15)$$

We defined m as the probability of mutation per replication. A different definition is often employed, however: m' is the probability of mutation per viral chromosome per replication. Since the viral chromosome increases two-fold during replication, the average number present during replication needs to be determined. The average of a continuous function is $\int f(x)\,dx / \int dx$. The relevant function here is $f(x) = 2^x$ where x is the number of replications. The average will therefore be $\int_0^1 2^x dx / \int_0^1 dx = 1/\ln 2$. Thus m' is m per $(1/\ln 2)$ viral chromosomes, or $m' = m \ln 2$.

A number of alternative methods which depend upon fluctuations in the numbers of mutants in replicate cultures are available for calculating mutation rates. These methods make use of actual mutant counts instead of merely the presence or absence of mutants. If we let r be the average number of *mutants* per culture among C cultures, then the average number of *mutations* per culture (M) was shown by Luria and Delbruck (1943) to be given by

$$r = M \ln (CM). \qquad (5\text{-}16)$$

This equation must be solved for M by interpolation. The mutation rate m is then obtained from equation (5-12). This formulation has the advantage over equation (5-13) of using more of the data available in an experiment, but it has the disadvantages of sensitivity both to selection artifacts and to jackpot cultures which may produce excessively large values of r. The latter difficulty may be overcome by using not r, but r_0, the *median* number of mutants among C cultures, expressed by Lea and Coulson (1949) as

$$(r_0/M) - \ln M = 1.24. \qquad (5\text{-}17)$$

This expression must also be solved for M by interpolation, M then being used to calculate m. The reader who is interested in the accuracy with which equations (5-14) through (5-17) may be applied is urged to consult the paper of Lea and Coulson, especially their Figure 3.

The remarks of this section have been set in the context of viral replication, but of course apply equally well to cellular replication. However, cells frequently contain two or more chromosome sets, each of which may contain more than one replication fork; as a result, determining the number of copies per cell of a particular gene can be difficult. In addition, a very significant interval frequently occurs between the appearance of a mutant chromosome in a cell, and the conversion of the cell to a mutant phenotype; this "phenotypic lag" can result in underestimated mutation rates. Furthermore, in at least some systems, mutation rates are proportional to division rates under some conditions, but are independent of division rates under other conditions (Kubitschek, 1960; Kubitschek and Bendigkeit 1964a, b; Kubitschek and Gustafson, 1964; Bendigkeit, 1966). Relationships of this type are most conveniently studied when the cellular growth rate is nutritionally controlled, and when a constant cell density is maintained by an automatic dilution process; the appropriate apparatus is usually called a chemostat. So long as they remain a small minority, mutants accumulate linearly with time in a chemostat. Both spontaneous and some chemically induced mutations accumulate at rates proportional to division rates in cultures whose growth is limited by the availability of glucose. In cultures whose growth is limited by the availability of tryptophan, however, mutation rates tend to be independent of division rates. This difference has been suggested to reflect different rates of repair of premutational lesions, or different rates of DNA synthesis, under the two growth conditions.

MUTATIONAL SPECTRA

When a large collection of point mutants is obtained and placed into sites by the methods of fine-scale mapping, the result is called a mutational spectrum. Truly extensive spectra have thus far only been obtained for the T4rII region, in which a large fraction of the sites which can be identified with moderate ease have in fact been discovered. In addition to their usefulness for determining the detailed geometry of the gene, mutational spectra are also useful for comparing the specificities of spontaneous and induced mutation. Even related mutagens tend to produce distinguishable spectra, for reasons which, however, remain obscure. The mutational spectrum of spontaneous T4rII mutants (Benzer, 1961) is shown in Figure 5-5. In this map, the sites are placed arbitrarily at equal intervals along the map. Segments defined by the ends of deletions (see Figure 4-5) are delineated by interruptions in the map, and are correctly ordered. The sites within the segments are usually not ordered, and are therefore placed arbitrarily. Each occurrence of a mutation at a given site is represented by a box.

The first aspect of the spectrum which impresses the eye is the great diversity of mutation rates at different sites. A total of 1609 mutants are distributed among 250 sites. However, 815 of these 1609 mutants, or just over half, fall into just two sites ($r17$, with 517; and $r131$, with 298). Many of the sites, on the other hand, are represented by single occurrences. The sites with large numbers of mutants are known as "hot spots." When a similar spectrum was obtained in the closely related bacteriophage T6, hot spots were also observed at positions corresponding to $r131$ and $r117$. The latter, however, was some four times "cooler" in T6 than in T4, suggesting that site-specific mutation rates are influenced by nearby bases.

How many sites have in fact been discovered? Two aspects of this question will be considered in order. The first problem is to estimate the total number of sites, and the second problem is to determine just how many of the mapped mutants actually identify sites.

Two methods have been used to arrive at the number of sites in the two rII genes, one stochastic and one which may be loosely termed chemical. Both Benzer (1961a) and Krieg (1963a) attempted to predict the total number of sites from the distribution of mutants among sites by assuming that the sites with one and two mutants fitted a Poisson distribution. Thus, $P(2)/P(1) = m/2$, where m is the average number of mutants per site; $m = M/S$ where M is the total number of mutants and S is the total number of sites, including the class represented by $P(0)$. This calculation predicted a total of 378 sites, of which 66 percent were already discovered. The calculation greatly underestimates the total number of sites, however, because of the very obviously nonrandom distribution of mutants among sites.

The second method for counting sites depends upon quantitative mapping. The total map length of the rII region, obtained by adding up intervals of less than R = 1 percent, is 16.86 (Edgar et al., 1962). Application of Figure 4-4 to these data (using the relationship that values below R = 1 percent correspond uniformly to 104 base pairs per 1 percent recombination) indicates a total of 1740 base pairs within the mapped regions. The actual total is probably somewhat higher, however, since regions at the extreme ends of the two genes may not have been included, and because of residual negative interference; a figure of 2000 base pairs is probably more realistic.

A much more subtle difficulty arises when one considers the requirements for discovering a site in terms of the architecture of mutational lesions. A mutation which consists of a base pair substitution clearly identifies a site. However, a mutation which consists of deleted or added base pairs will not in general identify a site which corresponds to a particular base pair in the wild type chromosome. (Such mutations do of course identify *map* sites, by definition.) Certainly an addition mutant should be capable of recombining at a finite rate with any base pair substitution mutant. A deletion mutant in which a single base pair has disappeared does define a site, but larger deletions do not do so unequivocally. It is fairly easy at present to identify mutants containing altered numbers of base pairs, but the actual number involved, whether they are added or deleted, can only rarely be determined. A large fraction of the sites appearing in Figure 5-5 contain mutants of this very type. Approximately 86 percent of rII point mutants appear to involve changes in the number of base pairs (Freese, 1959a; see also Chapter 15). Making a correction for the fact that the two most

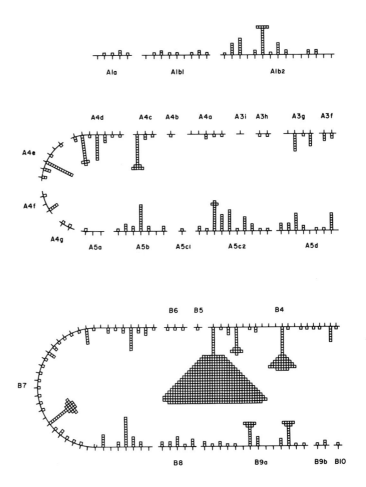

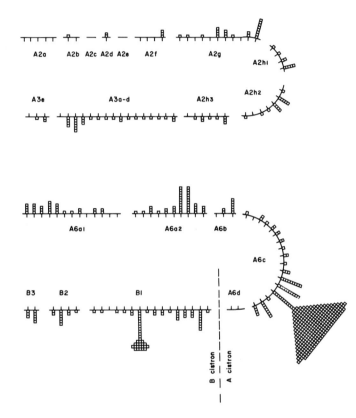

prominent hot spots contain mutants of this type, it is possible to calculate that only about 71 sites of the base pair substitution type were discovered. This is less than 4 percent of the estimated number of 2000 sites in the region.

Since data from other sources indicate that base pair substitutions are far from rare events, it would appear that many are simply not detected. Several factors probably conspire to conceal these mutants. One factor is codon degeneracy. Figure 5-6 shows the codon catalogue; inspection of the figure reveals that about 24 percent of the 549 possible base substitutions from amino acid codons fail to produce an amino acid substitution. Another factor is the plasticity of polypeptide composition. For example, extensive analyses of amino acid substitutions detected among related species of hemoglobins reveal that there are very few positions in the chain which can never be substituted, whereas chemically related amino acids may frequently be substituted one for the other in the chain (Perutz, Kendrew, and Watson, 1965). At the time of their analysis, only 9 of more than 140 residues were still observed to be invariant. In addition, elegant and detailed structural analyses of human hemoglobins (Perutz et al., 1968; Perutz and Lehmann, 1968) suggest that defective molecules will result mainly from amino acid substitutions at the substrate binding sites, at intermolecular binding sites, and at positions within the interior of the molecule (when polar residues are introduced); other substitutions will frequently be innocuous. A third factor tending to decrease the probability of scoring a mutational lesion is the frequent ability of organisms to make do with only a fraction of the activity of a particular gene. This may be possible because of regulatory phenomena, or simply because some gene functions can be partially reduced without producing the mutant phenotype, at least under laboratory conditions. This kind of behavior is characteristic of the rII region itself: considerable damage must accrue to the A or the B cistron in order to inactivate the ability of the virus to grow on K(λ) strains and therefore produce a mappable mutant. A considerable number of T4r mutants do in fact grow fairly well on K(λ), but still map within the rII region. They cannot be mapped into sites by the methods of fine-scale mapping, however, because of their great "leakiness."

Base pair substitutions frequently produce, not amino acid substitutions, but "CT" (chain termination) codons (Figure 5-6). These are the amber (UAG), ochre (UAA), and UGA codons, which terminate polypeptide synthesis. Because of their drastic effects, they are generally easily detected when produced by mutation. The spectrum shown in Figure 5-5 contains a considerable number of such sites. Calculations based upon the data of Benzer and Champe (1961), Brenner, Stretton, and Kaplan (1965), Brenner, Barnett, Katz, and Crick (1967), and Barnett et al. (1967) suggest that about 44 of the estimated 71 base pair substitution sites produce CT codons. The number of amino acid substitution mutants discovered then becomes 1.4 percent, a truly minute fraction of the estimated total number of sites.

One way in which the total number of identifiable sites might be sharply

2nd POSITION

		U	C	A	G	
1st POSITION	U	PHE PHE LEU LEU	SER SER SER SER	TYR TYR CT CT	CYS CYS CT TRY	U C A G
	C	LEU LEU LEU LEU	PRO PRO PRO PRO	HIS HIS GLN GLN	ARG ARG ARG ARG	U C A G
	A	ILU ILU ILU MET	THR THR THR THR	ASN ASN LYS LYS	SER SER ARG ARG	U C A G
	G	VAL VAL VAL VAL	ALA ALA ALA ALA	ASP ASP GLU GLU	GLY GLY GLY GLY	U C A G

3rd POSITION

FIGURE 5-6. *The codon catalogue. CT = chain terminating codon.*

increased, would be to impose restrictions on the functioning of the gene or its product. Lesser damages might then become of greater importance. This has already been done in the T4rII region by looking among leaky rII mutants for temperature-sensitive mutants, which can be mapped at 42°, but not at lower temperatures. Nineteen new sites were easily discovered, and many more sites could probably have been discovered if somewhat more restrictive conditions could have been devised. One way to approach this problem was suggested by the appearance of many of the temperature-sensitive mutants when plated on B cells: they tended to be semi-r in appearance. Against the semi-r background, full-r secondary mutants still stand out well. Many mutations which would themselves produce a very leaky rII, or even wild phenotype, are nevertheless fully mutant in the semi-r background (Figure 5-7). Using this approach, numerous new rII sites have recently been detected (Koch and Drake, unpublished).

Although the problem of mutant detection has been discussed from the standpoint of forward mutation, the principles apply equally to reversion. Thus a base pair substitution which results in a deleterious amino acid substitution, may be reversed by the substitution of either the original base pair, or a base pair which produces a nonwild type but nevertheless acceptable amino acid substitution. Numerous examples of events of this type will be discussed in future chapters.

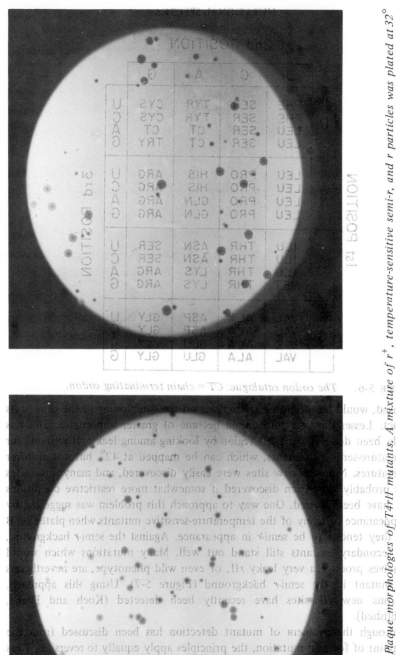

FIGURE 5-7. *Plaque-morphologies of T4rII mutants. A mixture of r^+, temperature-sensitive semi-r, and r particles was plated at 32° (left) and at 37°, (right). The temperature-sensitive mutant is indistinguishable from the ordinary mutant at 37° but it exhibits an intermediate phenotype at 32°. (Photos by R. Koch and J. Sprague.)*

THE ROLE OF DNA COMPOSITION

A mutation rate will reflect not only the probability of detection, but also the probability of repair of premutational lesions, as well as the intrinsic mutability of the genetic material itself. While forward mutation rates tend to cluster well within a factor of 10 above and below the average, reversion rates, and forward mutation rates at specific sites, are observed to vary over a truly fantastic range. Different T4rII mutants, for instance, exhibit reversion rates from as large as 10^{-2} per replication, to at least as low as 10^{-12}; the intervening 10 decades are well filled with observed mutation rates. Table 5-1 shows the reversion rates of a number of T4rII mutants; values larger than 10^{-4} were not recorded because such mutants are generally discarded, being too unstable to be mapped with any accuracy.

TABLE 5-1

REVERSION RATES OF T4rII MUTANTS

Mutant	Reversion Rate	Mutational Lesion
SM67	$\sim 5 \times 10^{-11}$	Frameshift
SM1	5×10^{-11}	Probably TA from GC
UV1	5×10^{-10}	AT from GC
UV237	1×10^{-9}	AT from GC
UV248	1×10^{-8}	AT from GC
UV55	6×10^{-8}	GC from ?
UV199	8×10^{-8}	AT from GC
SM32	2×10^{-7}	Frameshift
UV34	6×10^{-7}	Frameshift
UV102	1×10^{-6}	AT from GC
SM64	6×10^{-6}	GC from ?
SM37	3×10^{-5}	Probably TA from GC
SM23	1×10^{-4}	Probably TA from GC

The UV mutants are described in Drake (1963b) and the SM mutants in Drake and McGuire (1967a). Reversion rates were estimated by dividing the frequency of revertants in high-titer, nonjackpot stocks by 20. "Frameshift" indicates the loss or gain of base pairs.

Because of the great spread of site-specific mutation rates, it has long been common practice to assume that the mutation rate of a given site depends not only upon the base pair configuration at that site, but also upon the nature of nearby base pairs (see, for instance, Benzer and Freese, 1958, and Benzer, 1961a). Some grounds for this conclusion are available from a comparison of mutation rates of "nonsense" mutations at different locations within a cistron. Nonsense mutants contain the codons UAA, UAG, or UGA (in the language of the messenger RNA), and can only arise or revert by events involving one of the three corresponding DNA base pairs. Reversion rates of UAG ("Subset I") mutants were observed by Benzer and Champe (1961) to vary over a 47-fold

TABLE 5-2

T4rII NONSENSE MUTANTS ARISING AT SPECIFIC SITES

UAG Sites	Occurrences	UAA Sites	Occurrences
C204, HE122, X417 HD231, N11	1-3	HF220, HF240, HF219, HF245, HF208, X191, HD147, N21, UV375, 375, AP53, N31	1-3
N34, S172, S99, X237, HB129	8-14	N7, N17, HE267, N12, N24	4-6
N19, HB74, HB232, HB118, AP164	15-28	X20, 360, N55	9-19
EM84, S116, S24, N97	29-44		

Mutants were induced in extracellular T4 particles by hydroxylamine (Brenner, Stretton, and Kaplan, 1965).

range, and by Brenner, Stretton, and Kaplan (1965) to vary over a 10-fold range; variation over a 20-fold range was also observed in the case of UAA mutants. The significance of these variations is blunted by the fact that nonsense mutants can revert to codons representing a variety of different amino acids, of which a variable proportion may be competent to restore function. In a few instances, however, the amino acid contents of revertants have been characterized. The data of Stretton, Kaplan, and Brenner (1966) reveal a 3.3-fold variation in the rate of the spontaneous mutation UAG → UApy (py = pyrimidine), and a 9-fold variation in the rate of the 2-aminopurine-induced mutation UAG → UGG. Similarly, Table 5-2 shows the relative frequencies with which UAA and UAG mutants are induced in the forward direction by hydroxylamine: certain sites are obviously more highly mutable than others. (The UAG mutants arose from both CAG and UGG codons; the UAA mutants arose only from CAA codons.)

While consistent with the hypothesis, these observations do not prove the existence of neighboring base effects on mutation rates. Different regions within a cistron may display different mutabilities, for instance, because of the summation of factors which are essentially extrinsic to the cistron, such as the direction of transcription and the nearness to points where DNA synthesis is initiated. There clearly remains a need for an experimental analysis of site-specific mutation rates modified by nearby base pair substitutions.

In addition to modifying effects of neighboring bases by mechanisms which can only be guessed at present, it seems likely that mutation rates should depend upon the local relative AT content of DNA. AT-rich regions are expected to exhibit lower melting points, and therefore higher chemical reactivities as the groups normally involved in hydrogen bonding are more frequently exposed even at biological temperatures. No experimental attack on this problem has yet been reported, but some excellent material exists for study: in bacteriophage λ, several regions with distinctly different GC frequencies have been detected (Skalka, Burgi, and Hershey, 1968; Inman, 1967), and mutants whose reversion could be studied are liberally available.

Base pair frequency and sequence is of course not a complete description of DNA structure, since the DNAs of a great many organisms are modified after synthesis by the methylation of occasional bases, and in the rather special case of the T-even bacteriophages, by the glucosylation of the HMC residues. The effects of these modifications upon mutation rates remain unexplored, although methods are now available for growing organisms with unglucosylated or with unmethylated DNA (Gefter et al., 1966; see also the section on host-controlled variation in Chapter 19).

COMPARATIVE MUTATION RATES

The great differences in genome size, generation time, and physiology which are observed among organisms suggest that similarly marked differences in mutation rates might also be observed. For purposes of comparison, any of three expressions of mutation rate might be useful: the rate per base pair replication, or per gene replication, or per total organism per generation. Only forward mutation rates should be compared, however, so that a large number of events can be averaged. Unfortunately, accurate measurements of forward mutation rates under conditions which lead to the detection of an appreciable fraction of all heritable chromosomal alterations, are not often encountered. The scoring procedure, for instance, often systematically discards leaky mutants. In other cases, it is not clear whether the mutant phenotype results from all types of inactivation of the gene, or only from certain modifications.

We have already seen that the mutation rate for bacteriophage T2r mutants is close to 5×10^{-5} per replication (Luria, 1951). Furthermore, if a stock of T4 is grown from about 10^2 to about 10^{11} particles, it is usually observed to contain about 10^{-3} r mutants. Application of equation (5-9) again reveals a mutation rate of about 5×10^{-5} per replication. The great majority of these r mutants fall into three genes by complementation tests. (The remainder localize in a number of other genes, but their frequency is very low, suggesting that only a few lesions in these genes are scored.) The average mutation rate for these three genes is therefore about 2×10^{-5} per gene per replication. A very similar mutation rate has been observed at the cI cistron of bacteriophage λ (Dove, 1968).

Corresponding figures for mutation in bacteria are available from the *E. coli* L-arabinose system and the *Salmonella typhimurium* histidine system. Mutations in the arabinose gene cluster are detected when they occur in any of five genes: repressor gene, permease gene, and genes A, B, and C (Englesberg et al., 1962). The data of Weinberg and Boyer (1965), together with equation (5-9), reveal an overall mutation rate of roughly 2×10^{-6} per replication. The average rate per gene per replication is therefore about 4×10^{-7}. The mutation rate to *his⁻*, measured by fluctuation tests using equations (5-16) and (5-17) (Lieb, 1951), is about 2×10^{-6} per cell division. Since there are ten genes in this system (Ames and Hartman, 1963), the average rate per gene per cell division is about 2×10^{-7}. Both *ara* and *his* rates are probably somewhat underestimated because the newly arisen mutants are not efficiently counted by the screening systems which were employed. Nevertheless, these rates are markedly lower than the corresponding T4 and λ rates.

A particularly striking example of the ability of an organism to determine its own mutation rate is seen in the temperate bacteriophage λ (Dove, 1968). When multiplying as a prophage, most of its genes are simply elements of the host chromosome; when multiplying lytically, the virus establishes its own replication apparatus. Its mutation rate as a prophage is observed to be from 20 to 100 times lower than as an independent entity, in good agreement with the relative mutation rates in bacteriophages and bacteria.

Numerous data in higher forms indicate an average mutation rate per gene *per gamete* of about 10^{-5}. The coat color genes of the mouse seem to be particularly suitable for measuring forward mutation at specific loci. In this case the average of five mutation rates was 0.9×10^{-5} per locus per gamete, the range being $0.4 - 3 \times 10^{-5}$ (Schlager and Dickie, 1967). Dobzhansky (1962) has tabulated human mutant frequencies; the average for six dominant and nine recessive genes was 2.8×10^{-5}, the range being $0.4 - 4 \times 10^{-5}$. However, 10^{-5} is probably a large overestimate of the mutation rate per gene, since it necessarily represents measurements performed on the tail of the distribution of mutation rates (Stevenson and Kerr, 1967): mutation frequencies ten or one hundred times smaller would tend not to be recorded, except in the anecdotal form of single occurrences. For example, Cockayne (1933) has described a large number of inherited diseases of the skin and its appendages, many of which were observed within only a single pedigree. Furthermore, the human haploid genome contains sufficient unique DNA sequences to specify well over a million genes, a number which is incompatible with an average rate of mutation to drastically altered phenotypes in the neighborhood of 10^{-5} per gamete.

Much of the available data on comparative mutation rates is summarized in Table 5-3, adapted from Drake (1969). One of the most striking results is that procaryotes appear to maintain a constant mutation rate per organism per replication of approximately 0.2 percent. (However, it is not yet clear what trends will appear when more extensive data becomes available for eucaryotes.) All of the mutation rates listed are impressively large, especially when one considers that many mutations which are not scored under laboratory conditions are certainly occurring in addition to those which comprise the measured rate; it seems reasonable that many of these additional mutations are quite decisively scored by natural selection. In addition, only a fraction of all newly arisen mutations are eliminated by selection in the same generation; the cumulative

TABLE 5-3

COMPARATIVE FORWARD MUTATION RATES

Organism	Base Pairs per Genome	Mutation Rate per Base Pair Replication	Total Mutation Rate
Bacteriophage λ	4.8×10^4	2.4×10^{-8}	1.2×10^{-3}
Bacteriophage T4	1.8×10^5	1.7×10^{-8}	3.0×10^{-3}
Salmonella typhimurium	4.5×10^6	$2. \times 10^{-10}$	0.9×10^{-3}
Escherichia coli	4.5×10^6	$4. \times 10^{-10}$	1.8×10^{-3}
Neurospora crassa	4.5×10^7	0.7×10^{-11}	2.9×10^{-4}
Drosophila melanogaster	2.0×10^8	$7. \times 10^{-11}$	0.8
Man	2.0×10^9	?	?

For all entries through *N. crassa*, mutants arise predominantly during chromosomal replication (Drake, 1966b), and the mutation rate was calculated from equation (5-9). Where necessary, an average amino acid was assumed to weigh 110 daltons, as calculated from average amino acid frequencies among *E. coli* proteins (Seuoka, 1961), and a gene was assumed to contain 1000 base pairs, corresponding to a polypeptide weighing 36,700 daltons. Data for bacteriophage λ were obtained from Dove (1968); the mutation rate was calculated for the *c*I cistron, whose protein weighs about 30,000 daltons. The T4 mutation rate was calculated for the *r*II cistrons (0.6 of the *r* mutants), which are estimated to contain about 1770 base pairs (Stahl, Edgar, and Steinberg, 1964; Edgar et al., 1962). The bacterial data are collected in the text and in Cairns (1963). The mutation rate for the *ad3* locus of *N. crassa* was calculated from the mutant frequency in a growth experiment encompassing about 60 generations (Brockman and de Serres, 1963; Horowitz and MacLeod, 1960; de Serres, personal communication).

Mutations in *Drosophila* arise both during and independently of cell division (Muller, 1959) An estimate of the total mutation rate per sexual generation per haploid set of chromosomes (Mukai, 1964) is 0.4, and an approximate mutation rate per base pair (Rudkin, 1963) was calculated by assuming that all mutations arise during cell replication, and that an average of 40 cell generations separate the fertilized egg from the progeny gametes. Note, however, that different isolates from nature may show as much as 20-fold different, spontaneous mutation rates (Plough, 1941). The human genome (Davidson, Leslie, and White, 1951) contains about 40 percent of redundant species of nucleotide sequences (Britten and Kohne, 1968), whose contribution to the mutational target size is unclear.

mutant frequency (the mutational load; see Muller, 1950) is probably several times larger than the raw mutation rate. This is particularly likely if only a small fraction of the genome is ever used in a normal generation, as appears to be the case in *E. coli* (Kennell, 1968). It is therefore likely that most individuals of highly advanced species carry many deleterious mutations.

Forward mutation rates are of course the sum of a variety of types of molecular events, which will be discussed in later chapters; these include base pair substitutions, base pair additions and deletions, and on a larger scale, chromosome rearrangements of various types. The large differences in mutation rates per locus observed in different organisms therefore presumably represent large differences in the frequencies of most or all types of lesions. This supposition is supported by a comparison of the fraction of spontaneous mutations which consist of additions and deletions of small numbers of base pairs (see Chapter 12): about 80 percent of T4rII mutants, about 20 percent to 40 percent of *S. typhimurium his* mutants (Whitfield, Martin, and Ames, 1966; Margolies and Goldberger, 1968), about 15 percent of *N. crassa ad3* mutants (de Serres, personal communication), an indeterminate but significant fraction of *E. coli lac* mutants (Malamy, 1966), but at most only a few percent of *Schizosaccharomyces pombe ad6* and *ad7* mutants (Loprieno et al., 1969). The forces which act upon an organism to determine its mutation rate must therefore simultaneously affect a number of different processes.

These forces are largely unevaluated; the action of natural selection in determining mutation rates has, however, been given preliminary theoretical consideration by several workers, including Kimura, Maruyama, and Crow (1963), Kimura (1967) and Levins (1967).

The approximately constant mutation rate per organism observed among the procaryotes (Table 5-3) suggests that these forces may be of a very general nature, presumably constituting a balance between the deleterious effects of the mutational load and the cost of reducing the load by improving accuracy in chromosome replication systems, or by improving repair systems. If this is so, then the very high mutation rate observed in *Drosophila* suggests that the equilibrium might be dangerously close to a species-lethal level, particularly since the contemporary environment may be becoming much more mutagenic than at most times in the past.

6. Collecting mutants: procedures and precautions

It is hardly possible to discuss mutation rates and mutational mechanisms without also discussing the methodology, and particularly the pitfalls, of scoring and collecting mutants. In this chapter we will consider methods for avoiding selection artifacts when measuring mutation frequencies, especially in iral systems, and then we will take up the problem of how to collect independently arising mutants in the most efficient manner.

SELECTION ARTIFACTS

Any process which perturbs the measurement of a mutant frequency without the investigator's knowledge can eventually make him look very foolish indeed. Since we are concerned with erroneous measurements, and since errors can arise in a great multiplicity of ways, it is not generally possible to make a coherent catalogue of avoidable dangers. Instead, criteria will be considered for demonstrating the absence of selection.

The standard control for the detection of selection is the reconstruction experiment. An artificial mixture of original and mutant organisms is constructed in proportions approximating those to be observed in a typical experiment. The mixture is then treated *identically* to the experimental virus populations. If the proportion of mutants in the control remains constant, then selection is considered to be absent from the system. For example, if the reversion of T4rII mutants is to be examined, then the control mixture would typically consist of a nonreverting rII mutant (a deletion), plus a small frequency of wild type particles.

There are two reasons why reconstruction controls occasionally fail to reveal selection when it is actually occurring. First, the investigator may fail to treat the control and the experimental populations really identically. This usually happens because of the lack of a sufficiently detailed understanding of all of the possible variables of the experiment. Second, the events which eventually produce a mutant may first produce intermediate states subject to strong selection, even though the final mutant is not itself subject to selection. Specific examples of such intermediate states are premutational lesions and mutational heterozygotes, both to be discussed in later chapters.

If the question is merely one of whether or not mutation is occurring at all, then it is often possible to formulate very simple criteria to rule out selection. For instance, if free virus particles are mutagenized, then it can readily be determined from equations (5-3) and (5-7) that

$$\frac{M(h)}{M(0)} = \left(1 + \frac{Kh}{M(0)}\right)e^{-h} \tag{6-1}$$

Here M is the absolute mutant concentration (mutants per unit volume), K is the proportionality constant between mutational and lethal hits, and h is the average number of lethal hits delivered. A plot of $\log \frac{M(h)}{M(0)}$ against h (Figure 6-1) exhibits a maximum at $h = 1 - \frac{M(0)}{K}$. Thus the maximum only appears when the mutation rate is significantly greater than the spontaneous background.

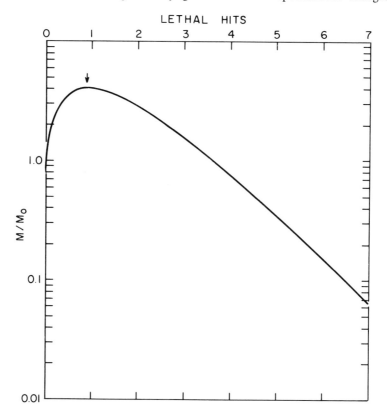

FIGURE 6-1.   *Relative number of mutations induced in free virus particles as a function of lethal hits delivered. For $K/M_0 = 10$ (see text), the maximum occurs at $h = 0.9$ (arrow).*

When it does appear, the absolute increase in the number of mutants is very difficult to explain in terms of selection artifacts.

No such simple method exists for eliminating the possibility of selection when mutants arise during virus replication. It is very common, for instance, to find that revertants of conditional lethal mutants grow better under permissive conditions than do the mutants themselves. It will be recalled that the method of measuring mutation rates by the *nonappearance of mutants* in a proportion of many identical cultures (equation [5-14]) is not sensitive to selection forces. This method, however, does require that the mutants, when they do occur, be detected with an efficiency equal to the efficiency with which unmutated particles are detected.

Many viruses exhibit cooperation among two or more inactivated particles infecting a single host cell, resulting in the production of a viable progeny. This phenomenon is called multiplicity reactivation, and is the cause of selection artifacts which seem to be peculiar to virus systems. If viruses inactivated with single-hit kinetics infect cells at an average multiplicity n, the probability that a cell will produce viable progeny should be $S = 1 - (1 - e^{-h})^n$; this is the

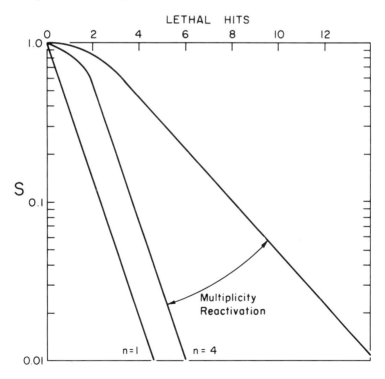

FIGURE 6-2. *Multiplicity reactivation. When viruses inactivated with single-hit kinetics ($n = 1$) infect cells at an average multiplicity of 4, the probability that a complex will produce viable progeny particles is given by the curve labeled $n = 4$ if multiplicity reactivation is absent, and by a flatter curve if it is present.*

multiple-hit equation (5-4). Multiplicity reactivation produces excess survivors (Figure 6-2). It frequently happens that a mutagen applied to free virus particles produces many lethal hits per mutational hit, and as a result, the total number of particles (active + inactivated) which must be plated to obtain a few mutants is very large. When this number approaches the number of plating cells, multiplicity reactivation can act to increase the apparent frequency of survivors, and thereby the apparent frequency of mutant plaques. Since the total number of surviving particles is usually scored at a higher dilution on a different set of plates, it is much less sensitive to multiplicity reactivation. Thus an apparent increase in mutants per survivor can be observed, even in the complete absence of actual mutation. Multiplicity reactivation has been recognized and analyzed as a complication in the induction of revertants of T4rII mutants by nitrous acid (Bautz-Freese and Freese, 1961).

Spurious data resulting from reactivation phenomena generally exhibit two characteristics which aid in their recognition. First, data from repeated experiments scatter quite strongly, and the scatter can often, with hindsight, be observed to reflect the ratio of total particles plated to total cells plated. Second, a plot of mutant frequency versus log of surviving fraction (equation [5-7]) curves upwards instead of remaining linear. When neither of these difficulties arises, and when reconstruction experiments are included, it is fairly safe to conclude that selection is absent.

A peculiarity of virus systems which may at times act to *obscure* the occurrence of mutations is phenotypic mixing, described in Chapter 3. It is easily eliminated by passaging a treated virus population once at a low average multiplicity.

COLLECTING MUTANTS OF INDEPENDENT ORIGIN

We will consider here the problem of collecting independently arising mutants. When a set of mutants is required for mapping purposes (such as for mapping a given cistron, or for estimating mutational specificity from a mutational spectrum), the independent origin of each mutant will usually be important. As we will see, independently arising mutants can often be obtained by fairly simple means.

If the mutants are induced in free virus particles, and are isolated directly, the problem is trivial: each mutant arises independently. It is only necessary to stipulate that the spontaneous background, which might contain sizable mutant clones, contributes little to the final mutant frequency. The recurrence of mutants at the same genetic site then clearly results from marked mutability at that site.

Suppose, however, that a set of mutants which arose by a replication-dependent process is gathered from a single lysate, and that a considerable number of them map at one site. Is this likely to have been the result of clonality (a jackpot stock), or rather the product of a highly mutable site? It is not generally possible to answer this question after the fact, except by obtaining another set of mutants from another lysate for comparison. However, it is a simple matter to demonstrate that very few mutants can be withdrawn from a given stock without running a high risk of coincidences (choosing mutants belonging to the same clone). During the final generation in the growth of a population, j mutations will occur, each producing a clone of size one; the total number of mutants produced is therefore j. During the preceding generation, $j/2$ clones will have arisen; but since each mutant duplicates once, a total of j mutants are again contributed to the final population. It is obvious by induction that each generation contributes an equal number of mutants to the final population. Therefore, if a population has increased in size by 2^i from the point when the first mutation was most likely to have occurred, considerably fewer than i mutations should be withdrawn from it in order to avoid the risk of a coincidence. As an example, consider a stock of T4 grown from a small inoculum to 10^{11} particles. Since the r mutation rate is about 5×10^{-5} per replication, $i = 22$, and no more than about half a dozen mutants can safely be harvested. The value of i can be increased by repeatedly recycling the stock, however, provided that inocula containing large numbers of mutants are transferred each time.

If a mutagen is used which induces a large increase of mutants over the spontaneous background within a single growth cycle, the treated complexes (infected cells) may be screened directly. If only a single mutant is isolated from each complex, the mutants will clearly have arisen independently. This procedure is essentially the same as treating free virus particles and harvesting the mutants directly. For example, if T4 is mutagenized during its growth in $E.$ $coli$, and the cells are plated before lysis, many will produce mottled plaques. (Mottled plaques appear in Figure 3-5, although their origin in that figure is not mutation.) One r mutant may then safely be isolated from each mottled plaque.

Another effective method for collecting mutants arising independently during DNA replication makes use of **antiselection**. This method is applicable whenever it is possible to select strongly *against* the desired mutants. Antiselection is applied during or after the growth of a stock to achieve a very low spontaneous background. In practice it generally entails growth of stocks under conditions which are nonpermissive for the particular type of mutant whose isolation is anticipated. Thus a stock grown at $42°$ would contain few temperature-sensitive mutants, and a stock of T-even bacteriophages grown in $K(\lambda)$ cells would contain few rII mutants. In each case, of course, the mutants which originated during the final cycle of growth would still be represented. Alternatively, antiselection may be applied to free virus particles. Thus h mutants in an h^+ population may be differentially adsorbed out by resistant cells, which may then be removed by centrifugation. After antiselection, the stock is allowed one, or at most a very few, additional cycles of growth without antiselection, in order to accumulate new mutants. Very large numbers of mutants may then be harvested from the stock.

EXCESSIVE MUTAGENESIS

It is frequently tempting to mutagenize a virus population to the point where it is very easy to collect large numbers of mutants from it. If the aim is to obtain one hundred mutants of a given gene, for instance, then mutagenization to the 1 percent level for that gene should result in only about one of the mutants collected actually being a double mutant. Since double mutants are usually fairly easily recognized in mapping studies, this seems like a reasonable procedure. Yet it is not.

The difficulty, of course, arises from the simultaneous induction of mutations in other genes at the same time, and also from the induction of undetected mutations in the primary gene. (While either of these possibilities may be obviated in principle by repeatedly backcrossing a given mutant to the wild type, this is obviously too difficult except in very special circumstances.) In a virus with about one hundred genes of nearly equal mutagenic susceptibility, where mutagenization has produced 1 percent of observable mutations in a given gene, 63 percent of the mutants would contain one or more additional mutational lesions in other genes. (Lethality of some of these secondary mutations might improve the situation, but only partially.)

Even more disturbing, especially when mutants are sought for fine-scale mapping and amino acid analysis rather than just for gene inactivation, is the possibility of undetected secondary lesions within the primary gene. It has already been noted that less than 5 percent of the potential sites in the T4rII region have been identified by means of base pair substitution mutations. Since we strongly suspect that the other sites are in fact mutable, but that mutations in them are not observed because of codon degeneracy and permissible amino acid substitutions, it follows that base pair substitutions may be introduced in this region 20 times more frequently than the corresponding mutations are detected. Chemical mutagenesis to the 1 percent level under these conditions, therefore, really means beyond the 20 percent level. The chemical analysis of amino acid substitutions in such mutants would frequently reveal multiple substitutions. Thus of 64 mutants of tobacco mosaic virus which contain amino acid exchanges in the coat protein, 13 have double exchanges and three have triple exchanges (Siegel, 1965).

ANOMALOUS MUTANTS

A complication which frequently arises when mutants are collected, and which often surprises the beginner, is the considerable proportion of mutants which for one reason or another turn out to behave in unexpected ways. Many mutants, the frequency depending upon the predominant type of mutational lesion and the requirements for structural integrity in the gene, are leaky. (This term refers to residual gene activity, and is conveniently defined operationally by comparison with a deletion mutant.) Furthermore, a mutant may easily appear not to be leaky by one criterion (for instance, by the screening procedure used to isolate it), but may exhibit considerable leakiness by another criterion. Many T4rII mutants, for instance, which exhibit large, sharp-edged plaques on B cells, nevertheless turn out to grow well on K(λ) cells (Edgar et al., 1962; Drake, 1966c; Drake & McGuire, 1967a). Sometimes special conditions can be devised to decrease the leakiness of mutants, such as growth at high temperatures or use of an alternate nonpermissive host cell. Most often, however, the mutants are discarded as being unsuitable for further use.

Another common type of mutant is unstable, not by virtue of leakiness, but because of an extremely high reversion rate. Whenever the reversion rate is greater than the supposed minimum recombination frequency between two adjacent mutants, the unstable mutant cannot be mapped into a specific site unless unusual procedures are adopted. Reversion may sometimes occur at extraordinarily high rates: rII mutants have been observed whose stocks always contain several percent of r^+ particles. The nature of these mutants remains an enigma, although the appealing possibility is often advanced that they contain duplications and revert by recombination.

One can imagine another type of mutant whose behavior would be extremely confusing. Consider a gene x whose gene product X was necessary either for the continued functioning of gene x, or else for the activation of gene x in progeny chromosomes. (The DNA methylating enzymes are conceivable candidates.) DNA replicated for any reason in the absence of the X substance would possess an inactive x gene, which would remain inactive until primed. Any attempt to map the resulting defect would produce progeny exclusively x^+ in both genotype and phenotype. Certain differences between bacteriophages T2 and T4 at one time appeared to behave somewhat like this. A cross between T2 and T4 produced progeny which were all as fully glucosylated as T4 (Streisinger and Weigle, 1956). It now seems likely, however, that DNA glucosylated only to the T2 level tends to be degraded in mixed infections (de Groot, 1967).

DANGEROUS COMPOUNDS

It would seem fairly obvious that mutagens are potentially dangerous compounds, even though the human body may largely detoxify many of them before they reach the germ cells. Most mutagens are in fact very toxic, so that workers either fear them to begin with, or else experience pain when making contact with them. The important exceptions to these generalities seem thus far to be the base analogues and the alkylating agents. For instance, Alderson (1965) observed that Drosophila which were fed a sugar solution for one day containing 0.4 percent ethyl methanesulfonate (EMS), showed 24 percent survival, and exhibited among their offspring a 70 percent incidence of sex-linked lethal mutations. Similarly, Benzer (1967) observed that flies fed 0.025 M (0.031 percent) EMS for 18 hours produced offspring containing about one sex-linked lethal mutation each. These data suggest that essentially nontoxic doses of certain compounds could produce a high incidence of mutations in man.

THE RARE EVENT PROBLEM

It is appropriate, when considering methodological problems, to emphasize a point which is rather obvious, but which is nevertheless frequently forgotten. Mutations are at best rare events, and mutations at a particular site are exceedingly rare events. Usually only the major reactions between a mutagen and DNA components tend to be well characterized; numerous rare reactions probably escape detection altogether. To correlate mutagenesis with one of the major reactions may at best be difficult; to eliminate the possibility that the mutations actually arise by infrequent and as yet uncharacterized reactions is nearly impossible. In later chapters we will see examples of very suggestive reactions, which probably turn out to be unimportant in mutagenesis after all. Heat and acid, for instance, produce both mutations and depurinations; but it now appears that depurination is rarely mutagenic, and that the mutations mostly arise by diverse reactions producing oxidative deamination of cytosine and other types of base degradation.

7. Mutations in viruses

Although this book is primarily concerned with mutational mechanisms, most readers will find it convenient to be introduced to the specific types of mutants which appear in viruses, especially since it will already have become obvious just how much mutation research has depended upon virus systems. Virus mutants may be broadly divided into two categories, namely, specific alterations of explicit functions or observable characters (such as endolysin production or plaque morphology), and conditional lethal mutations which inactivate functions which may not otherwise have been specifically identified. This division is roughly analgous to the separation of mutations in higher forms into visibles (such as eye colour alterations) and recessive lethals, the latter again being identified without knowledge of the specific function involved. Considerable crossing between these two categories occurs, however. For instance, some mutants identified as conditional lethals in a cistron mapping at about one o'clock on the T4 map (Figure 4-3), turn out to be located in the lysozyme (endolysin) gene, and can be grown under otherwise nonpermissive conditions simply by adding an exogenous source of lysozyme. Conversely, many mutations in genes near the *r*II region remain undiscovered, probably because these genes are largely dispensable under standard conditions of growth.

Large numbers of different types of mutations have been observed to occur in a great many viruses. For the most part, this chapter will discuss mutations affecting bacteriophages, since these are the mutations whose mechanisms of formation are best understood. Occasional examples involving other viruses will be included for comparison.

PLAQUE MORPHOLOGY MARKERS

It is hardly surprising that any virus which can form plaques can also sport plaque variants. Any reduction in burst size, which might occur mutationally in thousands to millions of different ways in a virus, would tend to produce a small plaque. While small-plaque variants are ubiquitous, they are usually unsuitable as genetic markers because revertants are strongly selected during the growth of stocks and within crosses. However, an array of markers such as *m* (minute), *w* (weak), and *tu* (turbid) appear in the T-even bacteriophages (T2, T4, and T6), and are sufficiently well behaved to have been used frequently before the advent of conditional lethal markers.

Many bacteriophages, including the genetically related T-even group, but not including, for instance, T1 or T3, form plaques with distinctly turbid margins. The cells in these regions are infected, but are **lysis-inhibited**. Lysis inhibition permits the extended synthesis of virus particles beyond the end of the normal latent period, resulting in greatly increased burst sizes, and is presumably very advantageous during the final stages of the infection of a population of cells. In the T-even bacteriophages, superinfection of an infected cell at any time between about 5 and 25 minutes (at 37°) after the primary infection, induces lysis inhibition (Doermann, 1948). The extent of lysis inhibition depends upon growth conditions, the number of superinfecting particles, and repeated superinfection (Bode, 1967). In bacteriophage ΦX174, on the other hand, a high cell density alone seems sufficient to induce lysis inhibition (Denhardt and Sinsheimer, 1965*a*).

A number of genes appear to cooperate in bacteriophage T4 to produce lysis inhibition. Mutations in any of these genes produce variants with large, sharp-edged plaques on *E. coli* B cells. These *r* mutants are located by mapping and complementation tests in at least three regions, *r*I, *r*II, *r*III, as shown in Figure 4-3. Many of the genes in the neighborhood of the *r*I gene seem to be dispensable under laboratory conditions, since they have not been identified by conditional lethal mutations. The *r*I gene is itself dispensable, although stocks of *r*I mutants usually exhibit low titers because of the absence of lysis inhibition. The *r*I gene is located close to the *e* (lysozyme) gene, in a region of the map which probably contains functions associated with virus release and cell destruction. Since lysozyme accumulates even in lysis-inhibited cells, the *r*I gene(s) presumably regulate the susceptibility of the cell membrane to lysozyme action; however, the biochemistry othe processes is unknown. The *r*III mutations, located at a considerable distance from the *r*I region, appear much more infrequently than do the *r*I mutations, and have not been studied in any detail. While *r*I mutations produce *r* plaques on all known host cells, *r*III mutations produce *r* plaques only on certain B strains, and produce r^+ plaques on K strains and many other B strains.

The two adjacent *r*II cistrons are located in a region of the map concerned with early alteration in the cell wall and the cell membrane. Their primary function is obscure, which is an odd circumstance when one considers that they are undoubtedly the most thoroughly mapped genes in any organism. They do, however, encode a polypeptide (McClain and Champe, 1967) which is produced in rather small amounts, and which is probably integrated into the cell membrane. Like *r*III mutants, *r*II mutants also produce r^+ plaques on K strains and on certain B strains. The most important characteristic of the *r*II mutants, however, is their inability to grow on λ-lysogens of *E. coli*, thus providing a basis

TABLE 7-1

PROPERTIES OF THE r MUTANTS OF BACTERIOPHAGE T4

Group	Plaque Morphology B	BB	K(λ)	Approximate Relative Spontaneous Frequency[1]
rI	r	r^+	r	0.3
rII	r	r^+	(0)	0.6
$rIII$[2]	r	r^+	r^+	0.1

[1] Refers to mutants arising during virus replication; values from Benzer (1957) and Drake (1966c,d).
[2] Consists mostly of leaky rII mutants.

for fine-scale mapping and reversion analysis (Benzer, 1955). The criteria for distinguishing rI, rII, and $rIII$ mutants are listed in Table 7-1, along with their relative spontaneous frequencies. Note that leaky rII mutants are usually indistinguishable from $rIII$ mutants in the absence of mapping experiments.

Another type of mutation producing large, clear plaques is peculiar to the temperate bacteriophages. Since these viruses may either lyse a cell or lysogenize it, they typically produce turbid plaques containing many lysogenic cells. Depending upon the conditions of plating, the turbidity may be uniform throughout the plaque, or may be concentrated into a central colony or even a central ring. Defects in any of several genes render the virus unable to lysogenize cells, thus converting it to a purely lytic mode of replication, and causing the production of **clear plaques**. These c mutants frequently fall into three cistrons (cI, cII, and $cIII$). The λcI cistron is of particular interest because it is the gene which produces the repressor which prevents the λ prophage from entering into a lytic cycle; this repressor has been isolated and characterized by Ptashne (1967). In addition, the cI gene in responsible for the inability of T4rII mutants to grow in λ lysogens (Tomizawa and Ogawa, 1967), and various mutant strains of λ differ greatly in their permissiveness for rII mutants (R. P. Freedman, personal communication).

A unique class of plaque morphology mutants has been used in genetic studies with bacteriophage T1. These mutants are **color variants** which are observed when dyes such as water blue and metrachrome yellow are included in the agar (Bresch, 1953). The wild type produces a characteristic blue halo around the plaque; c mutants produce tiny plaques lacking the blue halo; w mutants produce slightly small plaques with yellow halos; and the wz mutant produces pale violet plaques.

The local lesions produced on leaves by tobacco mosaic virus appear in a variety of genetically controlled sizes (Rappaport and Wildman, 1957; Wu, Hildebrandt, and Riker, 1960). A particularly useful type of mutation produces a local lesion response from a parental strain producing only systemic infections (Tsugita and Fraenkel-Conrat, 1960). Since genetic recombination in this virus is absent or extremely rare, however, its mutants are not as useful indices of mutation as are those in other viral systems.

Plaque morphology mutants have been employed in a variety of genetic studies with animal viruses. Among the large DNA viruses, vaccinia (a poxvirus) exhibits a number of plaque types, some of which can even be scored in the pock assay *in ovo* (Fenner and Comben, 1958). Among the RNA viruses of intermediate particle size, Newcastle disease virus (a myxovirus) produces many well-defined plaque types which also can readily be induced and reverted by mutagens (Granoff, 1961; Thiry, 1963). Small plaque variants are also common among the tiny RNA viruses. Animal virus plaque morphology mutants were in fact first observed in polioviruses (Vogt, Dulbecco, and Wenner, 1957). Small-plaque variants of mouse encepholomyocarditis virus owe their reduced size to inhibitors present in the agar (Takemoto and Liebhaber, 1961).

MUTATIONS AFFECTING HOST RANGE AND ADSORPTION

Most manifestations of host range among viruses depend upon the ability of the particles to adsorp to host cells. Viruses characteristically exhibit specific **ionic requirements** for adsorption, and may also show requirements for **organic cofactors**. Thus bacteriophage T4 adsorbs optimally in the presence of 0.1 M NaCl, and in addition, requires tryptophan as a cofactor. Mutations of T4 which abolish the requirement for tryptophan have been described (Anderson, 1948), as have mutations which introduce a requirement for calcium ions (Delbruck, 1948).

Bacterial strains resistant to bacteriophages can usually be selected directly; these strains can then be used to select virus mutants with **extended host range**. A strain of cells C resistant to a bacteriophage P is designated C/P, and the corresponding virus host range mutants are designated h. Patterns of host cell resistance and extended virus host range have been examined in many bacteriophage systems, but only those studied in detail and used in studies on genetic mechanisms need be mentioned here. The h mutants of T2, already mentioned at the beginning of Chapter 5, are anomalous in that they all arise at a given site in the gene, whereas h^+ "revertants" arise at a variety of sites (Streisinger and Franklin, 1956). The h mutants of T4, which arise at many sites (Edgar, 1958), may be selected by using *E. coli* K12/4 strains, but not by using B/4 strains. The h mutants of both T2 and T4 tend to be more sensitive to temperature and to chloroform than the corresponding wild types. Mutations outside of the h region can sometimes modify the host range phenotypes of these viruses (Streisinger, 1956; Baylor et al., 1957). The h regions of T2 and T4 are interesting in another way: although the two viruses show very extensive genetic homology, their wild type host range genes appear to be completely nonhomologous (Séchaud et al., 1965). As a result, h^{2+}/h^{4+} heterozygotes, which are easily scored as clear plaques on a mixture of B/2 and B/4 cells, consist exclusively of terminal redundancy heterozygotes. The same restrictions appear to exsist on the formation of these heterozygotes as exist on the formation of heterozygotes incorporating markers of extended deletions (see Chapter 3).

It should be noted that host range mutants may often be scored as plaque morphology mutants: if a mixture of T2h and T2h^+ particles is plated on a mixture of B and B/2 cells, the h particles will form clear plaques while the h^+ particles form turbid plaques.

An extensive series of host range mutants of bacteriophage T2 all form turbid plaques on B/2 cells (Baylor et al., 1957). They adsorb to B/2 cells poorly, and are often more sensitive to heat than are wild type particles. These mutants, called ht, scatter over more than half of the T2 map (Baylor, Hessler, and Baird, 1965).

A number of special host range systems very useful for mutational studies have been developed using the closely related bacteriophages ΦX174 and S13 (Tessman, Poddar, and Kumar, 1964). In some cases it was observed fortuitously that h mutants were unable to grow on wild type host cells; in other cases, additional host cell mutants were produced which could support the growth of h^+ but not of h particles. Either type of system permits the selection of mutations in both the forward and reverse directions, thus greatly facilitating studies on the specificity of various mutagens.

Certain strains of tobacco mosaic virus produce systemic infections in a given host, and also produce mutants exhibiting local lesions on that host. These mutants may, in turn, produce systemic infections in another host, and can thereby be grown into high-titer stocks. This "differential host" system (see Siegel, 1965) resembles host range systems to some extent, and has been very useful in mutational studies. Only a portion of these mutants affect the coat protein of the virus, however.

True host-range mutants, as opposed to mutants exhibiting altered virulence, are rare among animal viruses. Animal viruses have frequently been adapted to unusual hosts, but the process usually appears to require multiple mutational changes, and in addition, such alterations may well be largely disconnected from the adsorption process.

BACTERIOPHAGE SUB-STRAINS

A number of pairs of bacteriophages in common usage are extremely closely related, even to the point of common laboratory origin, but nevertheless exhibit significant differences. It has already been noted that ΦX174 and S13, small viruses with single-stranded DNA, are very closely related, so that experimental data for the two are more or less interchangeable (Tessman and Shleser, 1963).

Two varieties of T2 are in common usage, T2H and T2L (Hershey, 1946). Both derive from a common ancestor, but are distinguishable in careful quantitative serological tests. While the Hershey strain is unable to grow on either B/2H or B/2L, the Luria strain can grow on B/2H (Brenner, 1959).

Two varieties of T4 are also in common usage, T4D and T4B. Both again derive from a common ancestor (Luria, personal communication), but T4D has lost its requirement for tryptophan. The two strains have diverged sufficiently so that crosses between them segregate several types of plaques (Rutberg, 1969). Most of the *r*II mutants of T4 have been isolated from the Benzer strain, while most of the amber and temperature-sensitive mutants (to be described below) have been isolated from the Doermann strain. As a result, when *r*II and *am* or *ts* mutants are combined by genetic recombination, it is frequently advisable to perform backcrosses to achieve a uniform genetic background.

RESISTANCE TO INACTIVATION

An impressive array of agents has been tested over the years for the ability to inactivate viruses or to prevent their multiplication. Mutants with increased resistance to some of these agents have often been observed. Some of these mutants have been useful in genetic studies, and others have been useful in the analysis of virus structure and reproduction.

Different viruses exhibit very characteristic sensitivities to ultraviolet irradiation; bacteriophages, in fact, are often used as biological dosimeters, especially for internal monitoring. Mutants with increased UV sensitivity have occasionally been observed. Two mutants of T4, called *v* and *x*, produce viruses with 2.2-fold and 1.7-fold increased sensitivities, respectively, and also act independently of each other, so that the double mutant *v x* has a 4.4-fold increased sensitivity (Harm, 1963*a*). The *v* gene directs the repair of radiation lesions, for instance by the excision of pyrimidine dimers (Setlow, 1966). The *x* gene does not appear to direct a repair process, but is at least partially involved in genetic recombination (Harm, 1964). An interesting modification of UV sensitivity occurs in tobacco mosaic virus, where the particle sensitivity is determined in part by the coat protein (Seigel, Wildman and Ginoza, 1956). In this case, different UV sensitivities observed in intact particles disappear when free RNA is irradiated.

Heat-resistant mutants are easily obtained from viruses, and have often been used both as genetic markers and for discovering what genes control the primary structure of specific enzymes and capsid proteins. Furthermore, it is often possible to distinguish true revertants and pseudorevertants on the basis of their heat sensitivities, and thus ultimately to deduce mutational pathways. Sometimes, however, as in the case of bacteriophage T5, the frequency of particles exhibiting phenotypic but not genotypic heat resistance far exceeds the frequency of true heat-resistant mutants (Adams, 1953). It is very generally observed that heat sensitivities are strong functions of both the ionic and the organic environment, and that measurements of heat sensitivity tend to be poorly reproducible.

Many viruses are bounded by semipermeable membranes. If sodium chloride is added to suspensions of T-even bacteriophages to 3 M or higher, and the suspension is then poured into an excess of distilled water, the particles are inactivated and their heads are disrupted, releasing DNA into the medium (Anderson, Rappaport, and Muscatine, 1953). Mutants resistant to osmotic shock are easily obtained. The reversion of these o mutants to o^+ is also easily monitored because of the increased sensitivity of o mutants to nitrogen mustard (Brenner and Barnett, 1959): back-extrapolation of the two-factor inactivation curve $(S = f_1 e^{-k_1 t} + f_2 e^{-k_2 t}$; compare equation [5-3]) provides the frequency of o^+ mutants. T4o, but not the wild type, is shocked by sucrose (Cummings, 1964).

Few drugs specifically inhibit the reproduction of viruses, in the sense that a viral and not a host function is affected. One striking exception is the inhibition of the growth of T-even bacteriophages by acridine and its derivatives. Several of these compounds are powerful mutagens and interact strongly with DNA. Their inhibitory effects on the T-even group, however, seem to depend upon inhibition of maturation (DeMars et al., 1953). Mutants resistant to dye appear at two loci (Hessler, 1963; Edgar and Epstein, 1961; Piechowski and Susman, 1967). One locus, near the rII region (ac in T4, pr in T2), primarily affects the response of infected cells to acridine and proflavin; another locus (q in both T2 and T4), situated within gene 17, primarily affects the response of infected cells to quinacrine. The ac mutation affects the permeability of infected cells (Silver, 1965 and 1967). Both types of mutants may revert by outside suppressor mutations which are concentrated in the cluster of genes determining DNA synthesis (Hessler, Baylor, and Baird, 1967).

The multiplication of a variety of enteroviruses (such as polioviruses) is selectively inhibited by guanidine and by 2-(α-hydroxybenzyl)-benzimidazole, and mutants resistant to either are readily obtained (Eggers and Tamm, 1961; Melnick, Crowther, and Barrera-Oro, 1961).

LYSOZYME

Many bacteriophages direct the synthesis of lysozymes which effect the release of mature particles from infected cells. Mutants blocked in this function (e mutants in T4) can be supplemented with exogenous egg white lysozyme (Streisinger et al., 1961). Revertants can be selected on unsupplemented medium, and occur both intracistronically and in outside genes. The complete amino acid sequence of the T4 lysozyme is now available, making this an excellent system for the analysis of gene/protein relationships; the enzyme contains 164 amino acids, and weighs 18,635 daltons (Tsugita and Inouve, 1968).

GENERALIZED CONDITIONAL LETHAL MUTANTS

A mutant which is able to grow satisfactorily under one set of conditions, but not under another, is called a **conditional lethal mutant**. Thus T4rII, h, ac, and e are conditional lethals. The main advantage of conditional lethals as markers in haploid organisms arises from the ease with which rare wild type particles may be selected. However, all of the conditional lethals thus far discussed require special, marker-specific permissive conditions for growth. In contrast, many conditional lethal mutants exhibit alterations which are phenotypically repairable without reference to the particular function inactivated. Discovering the exact function affected may, in fact, prove to be a considerable task. Two broad classes of such generalized conditional lethal mutations have been characterized in viruses: suppressible mutations and temperature-sensitive mutations.

The mechanism of suppression will be examined in some detail in Chapter 16, while only the main properties of suppressible mutants will be discussed here. The suppressible mutants in most frequent use contain **nonsense codons** (Benzer and Champe, 1962a), that is, codons which specify no amino acid, but which cause premature termination of the polypeptide chain (Sarabhai, Stretton, and Brenner, 1964). The codons UAA, UAG, and UGA are chain-terminating, the UAA codon probably being used most frequently for natural chain termination. The properties of the UAG type of mutant were discovered before the codon was identified, and the name *amber* was assigned to the mutants suppressed by a particular strain of *E. coli* (strain CR63) (Epstein et al., 1963). (The first *amber* mutant was isolated by Harris *Bernstein,* and appropriately Anglicized.) During the analysis of the *amber* codon, the UAA was also recognized as nonsense, and was named *ochre* by Sydney Brenner in order to achieve an appropriate chromatic coherence. By the time the UGA codon was recognized, however, there were distinct signs that colorful names might proliferate excessively, and no assignment was published, although the term *opal* has had a limited private circulation.

The suppression of *amber* mutants is efficient, in the sense that the chain is propagated normally as frequently as 63 percent of the time (Kaplan, Stretton, and Brenner 1966). The amino acid which is inserted at the site of the UAG codon is different in different strains of permissive host cells: strain CR63 inserts serine, while other strains insert tyrosine or glutamine. However, since chemically induced *amber* mutations usually arise from tryptophan and glutamine codons, the protein produced by suppression often contains an amino acid different from the wild type. The suppressed protein may therefore be inferior to the wild type protein, and a net efficiency of suppression less than 63 percent may easily ensue. Strong suppressors now exist for UGA codons as well as for UAG codons (Sambrook, Farr, and Brenner, 1967), and UGA mutants will probably soon come to be used as extensively as are *amber* mutants. Suppressors of *ochre* mutants usually achieve chain propagation very inefficiently, thus sharply limiting the usefulness of *ochre* mutations as genetic markers.

A number of suppressible mutants, such as the *sus* mutants of bacteriophage λ (Campbell, 1961) and the "Subset I" mutants of the T4rII region (Benzer and Champe, 1961), have turned out to be *amber* mutants. In addition, *amber* mutations have now been found in an ever-increasing number of viruses, and are undoubtedly ubiquitous. In addition to the obvious method of picking and testing plaques, *amber* mutants may be efficiently isolated by plating viruses with the permissive cell strain, overlaying with soft agar containing the nonpermissive cell strain, and looking for small or turbid plaques.

The nomenclature for describing permissive cell strains, which are themselves mutants, is sometimes confusing to the beginner. The wild type is su^-, while the suppressor strain is su^+, or sometimes su_I^+, su_{II}^+, etc., to distinguish different strains.

At about the same time that *amber* (*am*) mutants were being dicovered in T4, temperature-sensitive (*ts*) mutants were also being discovered (Edgar and Lielausis, 1964). (Temperature-sensitive mutants had, however, been observed for decades in a variety of other organisms.) Viral *ts* mutants can easily be isolated by incubating plates at a permissive temperature for a few hours, switching the plated to a nonpermissive temperature, and then looking for retarded plaques. The most common temperature differential used to define a set of mutants is 25-32° (permissive)/41-43° (nonpermissive). However, there is in principal no compelling reason why any sufficiently large differential, in either direction, cannot be used to define a set of *ts* mutants. Thus cold-sensitive mutants have been observed in ΦX174 (Dowell, 1967), in T1 (Christensen and Saul, 1966), and in T4 (Scotti, 1968).

Temperature-sensitive mutants are widely believed to be **missense** mutants, that is, to contain amino acid substitutions rather than chain-terminating codons. While the experimental evidence in favor of this assumption is meager, there are no compelling reasons to seriously doubt it.

The relative advantages and disadvantages of *ts* and *amber* mutants as conditional lethals depend upon several general differences between the two. First, the *amber* lesion produces complete gene inactivation, whereas the *ts* lesion frequently produces a leaky mutant. Although insufficient to produce a visible plaque, an average burst size of 2 or more may well be achieved at the "nonpermissive" temperature. This type of behavior can seriously compromise both reversion and complementation tests with *ts* mutants. Second, it is possible to alternate between permissive and nonpermissive conditions during a single growth cycle with *ts* mutants, but not with *am* mutants. This property makes *ts* mutants particularly valuable for the analysis of steps in viral multiplication (Edgar and Epstein, 1964; Horiuchi, Lodish, and Zinder, 1966). However, as described in Chapter 16, certain drugs such as streptomycin and 5-fluorouracil can partially overcome the *amber* block, thus permitting some variation in permissiveness during a single growth cycle (Edlin, 1965). Third, *amber* (but not *ts*) mutants are sometimes polar. Polarity will be discussed in some detail in Chapter 18; it is sufficient here to note that polar mutants not only inactivate the gene in which they fall, but also reduce to various extents the activity of one or more adjacent genes. Polarity seems to be infrequent in viral systems, but has been detected both in T4 (Nakata and Stahl, 1967) and in S13 (Tessman, Kumar, and Tessman, 1967).

The *amber* and *ts* mutations have served admirably in the detection of a large fraction of the genes of bacteriophage T4 (Figure 4-3), and also of the genes in several other bacteriophages; the map of the minute bacteriophage S13 is shown

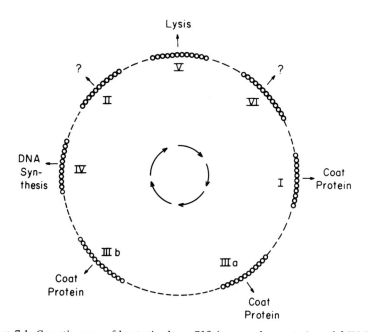

FIGURE 7-1. *Genetic map of bacteriophage S13 (a very close relative of ΦX174). The polarity of the chromosome is deduced from the polar behavior of amber mutations in gene IIIa (see Chapter 18) and from the fact that only one strand of the double-stranded replicative form of the virus is transcribed into messenger RNA (see Chapter 8). Gene sizes and intergene distances have not been determined and are arbitrarily assigned.*

in Figure 7-1 (Baker and Tessman, 1967; Tessman, Kumar, and Tessman, 1967). The degree of saturation of the T4 map may be judged from the fact that a recent collection of about 1000 new T4 *amber* mutants increased the number of known genes by only 9. Including the mutations detected by other criteria, about 80 genes are now identified in T4. The corresponding number in S13 is 7.

It is interesting to compare the number of genes identified by *cis-trans* tests with the number of genes expected on the basis of the nucleic acid contents of these two viruses. Taking base frequencies and amounts of DNA from data tabulated in Luria and Darnell (1967), and assuming an average molecular weight for an amino acid of 110, then T4 can code for about 6.6×10^6 daltons of protein, while S13 can code for about 2.0×10^5 daltons of protein. The average weight of the elementary polypeptides of many proteins from many species is about 40,000 daltons, while the average of five proteins of T4 (lysozyme, sheath, head, *r*IIA, and *r*IIB), for which fairly good estimates of molecular weight exist, is about 44,000 daltons. Using the figure of 40,000 daltons, the estimated gene number for T4 is 165; the corresponding figure for S13 is 5. It therefore appears that only about half of the genes of T4 have been discovered, whereas all of the genes of S13 are probably identified. The undetected genes in T4 are probably genes which are at least partially dispensable under common laboratory conditions. Examples of dispensable genes are the *r*II genes, as well as *wh, td,* and *cd* (Figure 4-3); lesions in the latter three genes, which control steps in pyrimidine biosynthesis, may reduce bursts sizes somewhat, but by no means to 0 (Hall and Tessman, 1967). A hunt for *amber* mutants which produce small plaques on nonpermissive host cells, rather than no plaques at all, might reveal a large number of additional genes.

8.   The taxonomy of mutational lesions

We will turn now from the genetics and physiology of viruses, to the details of the mutation process itself. It will be convenient to begin by considering the elementary types of mutational changes (Figure 8-1). These changes can be distinguished initially according to whether they produce "large" or "small" alterations in the DNA molecule, although, as we shall see, a certain amount of overlap does connect the two classes. The taxonomy which will be employed here derives primarily from the work of Benzer, Freese, Brenner, and Crick, along with their associates. Some of the terminology proceeded through a difficult developmental period, with the result that a few of the early papers appear out of register with later papers. However, the present system is consistent, both internally and with the great majority of the experimental results.

MICROLESIONS

Point mutations are broadly divided into two classes, base pair substitutions and frameshift mutations. Base pair substitutions affect only one base pair at a time, but frameshift mutations consist of deletions and additions of small numbers of base pairs. Although frameshift mutations can probably approach the "large" mutations in size, most are actually quite small, and also readily revert; it is therefore more convenient to consider them all as point mutations. It is only in the rather infrequent circumstance that large numbers of such mutants are mapped within a fraction of a cistron (Barnett et al., 1967), and hence that complex overlap patterns appear, that frameshift mutations reveal their multisite nature.

The twelve possible base pair substitutions were divided by Freese (1959c) into two groups, the transitions and the transversions. Transitions are mutations in which the purine-pyrimidine orientation is preserved, or more specifically, in which a purine is replaced by another purine, and/or a pyrimidine by another pyrimidine. Transversions are mutations in which the purine-pyrimidine orientation is reversed, a purine being replaced by a pyrimidine and/or a pyrimidine by a purine. This scheme is illustrated in Figure 8-2.

away, and already the doomsday
warnings are arriving, the fore-
boding accounts of a Russian
horde that will come sweeping
out of the East like Attila and
his Nuns.—*Red Smith in the
Boston Globe.*

SUBSTITUTION

"I can speak just as good
nglish as you," Gorbulove corrected in a merry voice.—*Seattle
Times.*

DELETION

"I have no fears that Mr.
Khruschev can contaminate the
American people," he said. "We
can take in stride the best brain-
washington he can offer."—
Hartford Courant

INSERTION

He charged the bus door
opened into a snowbank, causing
him to slip as he stepped out and
uɐɹ ɥɔᴉɥʍ 'snq ǝɥʇ ɥʇɐǝuǝq ꞁꞁɐɟ
over him.—*St. Paul Pioneer
Press.*

INVERSION

Tomorrow: "Give Baby Time
to Learn to Swallow Solid
Food."
etaoin-oshrdlucmfwypvbgkq
—*Youngstown (Ohio) Vindicator.*

NONSENSE

FIGURE 8-1.   *Typographical errors, illustrating various types of mutations
(Benzer, 1961b).*

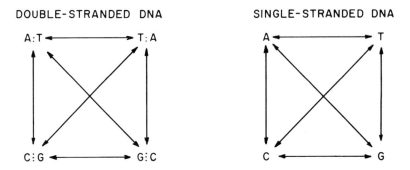

DIAGONALS = TRANSITIONS
HORIZONTALS & VERTICALS = TRANSVERSIONS

FIGURE 8-2.   *One dozen base pair substitutions.*

The classification of frameshift mutations depends upon the fact that messenger RNA is translated in units of three bases (= one codon) at a time. Once the direction and phase of translation are established by the initiation process, the triplet reading process continues until a chain-terminating codon is reached. If a base is inserted into the polynucleotide, the result will be to produce a reading frameshift to the left. If a base is deleted, the result will be to produce a reading frameshift to the right. This scheme is illustrated in Fig. 8-3.

Since the codon consists of three bases, the addition of one base produces the same frameshift as does the addition of four bases, or the deletion of two bases. Similarly, the deletion of one base is equivalent to the addition of two bases, either event shifting the reading frame to the right. In more general terms, frameshift mutants can be considered to fall into three classes:

$$+ 1 \pm 3n \text{ base pairs}$$
$$\pm 3n \text{ base pairs}$$
$$- 1 \pm 3n \text{ base pairs}$$

where n may assume any integral value, including 0. The middle class is said to contain mutants of sign (0). The other two classes are said to contain mutants either of sign (+), or of sign (−). Operationally, however, while mutants may easily be divided into groups of opposite signs, it is not generally possible to deduce which group contains (+ 1) mutants, and which group (− 1) mutants, from simple genetic measurements. When such mutants were first characterized, therefore, one class was arbitrarily called (+), although it was much later before a decision could be made as to whether these mutants were of the (+ 1) or of the (− 1) variety.

READING FRAME SHIFTS

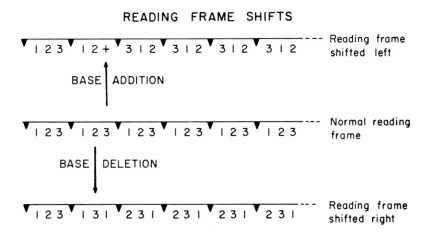

FIGURE 8-3.   *The multiple effects of single base additions and deletions on the translation of messenger RNA.*

Base pair substitutions produce characteristic effects upon proteins. When they occur within a gene, they produce either amino acid substitutions, or else chain-terminating codons (UAA, UAG, or UGA). Amino acid substitutions often produce leaky and/or temperature-sensitive mutants. They also frequently produce proteins which may be detected serologically, even in the complete absence of enzymatic activity, by the ability of the inactivated enzyme to remove from solution antibodies which could otherwise be scored by their ability to inactivate wild type enzyme; such mutants are said to produce cross-reacting material (CRM). The chain-terminating mutants are nearly always nonleaky, and are suppressed by specific strains of bacteria. As we shall see, many base pair substitution mutations can also be induced to revert by specific mutagens.

It was noted earlier that codon degeneracy and plasticity of protein composition may conspire to conceal a large majority of base pair substitution mutations. However, it is possible to imagine conditions in which codon degeneracy will be insufficient to conceal a mutation. Different codons for the same amino acid are frequently recognized by different species of transfer RNA (Weisblum, Benzer, and Holley, 1962; Kellogg et al., 1966; Söll et al., 1966). The relative amounts of synonomous species of transfer RNA may be quite different. For a protein stoichiometrically related to cell mass or virus burst size, a change of codons for the same amino acid would produce a phenotypic change whenever the new transfer RNA was present in insufficient amounts. In more general terms, whenever the differential rate of translation of synonomous codons produces a protein in insufficient quantities, an extremely leaky mutant could result. A preliminary attempt to detect different levels of synthesis of *E. coli* alkaline phosphatase when a particular residue was coded by different codons, was unsuccessful (Weigert et al., 1966; see also Anderson, 1969).

Frameshift mutations also produce characteristic effects upon proteins. From the mutational site onwards, each codon will be different, and as a result, each amino acid will be different within the limitations of codon degeneracy. However, a chain-terminating codon will usually be reached very soon. The net result is that frameshift mutants are virtually always nonleaky and CRM-less. They are generally not recovered when they occur in viral somatic proteins, whereas base pair substitutions leading to amino acid substitutions may frequently by recovered; *h* mutations, for instance, do not arise from frameshifts, but do arise from amino acid substitutions. However, frameshift mutations of sign (0) (insertions or deletions of multiples of three base pairs) might sometimes behave like base pair substitutions. Mutations of this type have not yet been recognized in virus systems, but could probably be detected by appropriate searches.

Another important difference between base pair substitution and frameshift mutations emerges when their effects are studied in diploid organisms. Frameshift mutations tend to be completely recessive, whereas base pair substitutions frequently exhibit partial dominance (Wills, 1968). Presumably, the mildly altered gene product resulting from a base pair substitution is more likely to interfere with the normal gene product than is the drastically altered gene product resulting from a frameshift mutation (see Chapter 18).

A special method has been developed in bacteriophage T4 to distinguish mutations with drastic effects from mutations with mild effects. A mutation of the *r*II region called *r*1589 is a deletion with unusual properties (Fig. 8-4). One end of the deletion falls in the A cistron, and the other end falls in the B cistron. The punctuation between the two cistrons has been deleted, producing a single compound cistron. Surprisingly, this cistron still possesses the B function in complementation tests, although the A function has been lost (Champe and Benzer, 1962*a*): the initial portion of the B cistron is dispensable for growth in λ-lysogens. When double mutants are constructed of *r*1589 plus a mutant in the remainder of the A cistron, the B function is sometimes inactivated (Crick et al., 1961; Benzer and Champe, 1962). Frameshift mutations and some deletions [presumably those of sign (+) or (−)] in the A cistron inactivate B function, as do chain-terminating base pair substitutions; base pair substitutions which produce amino acid substitutions do not inactivate the B function. Lesions which inactivate the B function are called nonsense mutations, while lesions which do not inactivate the B function are called missense mutations.

MACROLESIONS

Disorders of the DNA molecule involving large numbers of base pairs fall into three groups: deletions, duplications, and rearrangements. The last group contains changes which have not altered the number of base pairs, but have either inverted a segment of the chromosome, or else exchanged two different

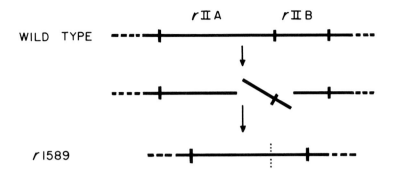

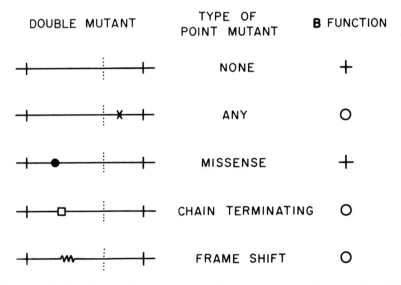

FIGURE 8-4. *A test for missense and nonsense mutations in the T4rIIA cistron. Top: the deletion r1589 generates a compound cistron. Bottom: the phenotypes which result when various rIIA point mutants are combined with r1589 reflect the molecular configuration of the point mutants.*

segments. In organisms with more than one chromosome, exchanges may also occur between different chromosomes.

It has already been noted that large frameshift mutations may resemble small deletions and duplications. A convenient criterion for separating small deletions from large frameshift mutations of the deletion type is the failure of the former to revert. Since duplications apparently do revert, however, no such simple and unambiguous criterion is available to separate them from large (addition) frameshift mutations.

Deletions are readily recognized, first by their inability to revert, and second by their failure to recombine with two or more point mutations. They may, in addition, inactivate two or more cistrons. Duplications, on the other hand, are more difficult to recognize. They may map as point mutations within a cistron, or may show anomalous mapping properties, depending upon what genetic material is duplicated, and where the duplicated material is inserted. Duplications should in principle be able to revert, and may in practice revert at very rapid rates. While large numbers of deletions have been detected in viruses, very few duplications have been recognized; perhaps some mutants, which in preliminary tests have been classified as containing point mutations, really contain duplications.

One of the most reliable tests for demonstrating the physical composition of a mutation which is suspected of being either a deletion of a duplication, is to measure its effects on the recombination of a pair of markers, one on either side of the mutation (Fig. 8-5). Deletions are expected to decrease the recombination frequency between such markers (marker shrinkage), and this expectation has been fulfilled, both with deletions within the T4rII region (Nomura and Benzer, 1961; Bode, 1963), and with a deletion of bacteriophage λ (Jordan, 1964). Marker shrinkage tests distinguish between true deletions and the substitution of stretches of random or unrelated base pair sequences, which might result, for instance, from exchange type rearrangements, or possibly even from inversions. Duplications, on the other hand, are expected to increase the recombination frequency of the outside markers (marker expansion). Whenever marker expansion or shrinkage is detected, however, control tests should be performed to ensure that the effect is not due to an alteration in the rate of recombination throughout the whole cistron or chromosome.

Nomura and Benzer (1961) suggested that deletions and duplications could be identified and measured with potentially great accuracy by extracting double-stranded DNA from mutant and wild type particles, mixing the two DNA species together, heating to denature, cooling under conditions allowing reannealing, and examining the hybrid molecules under the electron microscope for regions where single-stranded loops extend from double-stranded molecules. Recent experiments with bacteriophage λ have dramatically verified the potential of this approach (Davis and Davidson, 1968; Westmorland, Szybalski, and Ris, 1969): deletions and regions of nonhomology were both identified and mapped, and one deletion was shown to have a highly unusual structure.

Since deletions and duplications may change the net amount of nucleic acid in a virus, they may correspondingly alter the physical properties of the virus particles. The λ deletion $b2$ decreases the density of the virus by an amount corresponding to an 18 percent loss of DNA, while chemical measurements indicate a 17 percent loss (Kellenberger, Zichichi, and Weigle, 1961). However, attempts to detect decreased densities in particles of bacteriophage T4 carrying deletions in and beyond the rII region were uniformly unsuccessful (Nomura and

Marker Shrinkage Test:

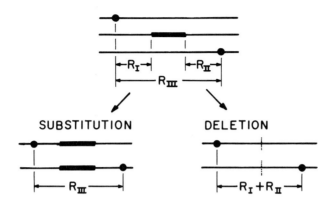

Marker Expansion Test:

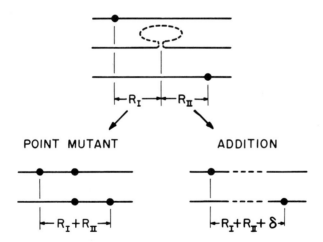

FIGURE 8-5. *Altered recombination frequencies between pairs of markers caused by centrally located macrolesions.*

Benzer, 1961). It is now clear that this result was unavoidable: terminal redundancy is simply increased in such mutants to compensate for the loss of deleted base pairs (Streisinger, Emrich, and Stahl, 1967). Additions, on the other hand, should decrease terminal redundancy in the T-even bacteriophages. A duplication or addition sufficient to exceed the length of the terminal redundancy would result in the production of a considerable fraction of inviable particles due to gene losses. Furthermore, the few viable particles produced would frequently be missing dispensable genes.

It seems likely that viable chromosome rearrangements in viruses would be recognized only by mapping studies sufficiently detailed to demonstrate changes in the relative positions of various markers. However, since many of the larger rearrangements might have ends falling in two or more cistrons, they might be sought initially among mutants showing loss of two or more functions. The discontinuities might even map as discrete sites, the intervening regions recombining normally because of the small size of the pairing region (Drake, 1967). Presumably as a result both of the difficulties of recognizing them, and of their infrequent occurrence in viable particles, chromosomal rearrangements are almost unknown in viruses. Only a few putative examples can be given.

An unusual mutant was detected by McFall and Stent (1958) in a stock of bacteriophage T2 which had experienced extensive P^{32} "suicide." The mutant appeared to contain two lesions, one in the rII region and the other probably in the very distant rI region (Fig. 4-3). In crosses with the wild type, however, segregants were produced which exhibited a variety of different linkage arrangements for the markers involved. The behavior of this mutant most nearly resembles what might be expected of an inversion. The DNA scissions required for its formation might well have appeared during the disintegration of P^{32} within the tightly packed virus head, and recombination with the wild type would be expected to produce new and unusual linkage relationships.

It has been pointed out by Muller, Carlson, and Schalet (1961) that "ultraminute" inversions of single base pairs are indistinguishable from transversions (for instance, $AT \rightarrow TA$). This suggestion requires that the transversions arise as a result of multiple polynucleotide interruptions, rather than as a result of bizarre mispairings (perhaps involving chemically altered bases). However, the processes by which transversions arise in bacteriophage T4 are single-hit and produce mutational heterozygotes, strongly suggesting a mispairing mechanism (Drake and McGuire, 1967a and b). In the absence of evidence that transversions may arise by mechanisms which also produce inversions, we will continue to include transversions in the class of base pair substitutions.

Rearrangements definitely do occur as a result of exchanges between the chromosomes of certain temperate viruses and the chromosomes of their bacterial hosts. The resulting hybrids are usually, but not always, defective; that is, they cannot complete a cycle of lytic growth unless an undamaged viral chromosome is also present. (However, defective viruses may persist indefinitely when replicating as prophages.) Since these hybrid viruses carry bacterial genes, they are capable of transduction. The exchanges occur in limited portions of the bacteriophage chromosome, although the exact location of the end points is variable (Campbell, 1959). Furthermore, the exchanges are not quantitatively reciprocal, the amount of bacterial DNA inserted being either more or less than the amount of viral DNA displaced. As a result, the defective particles have unusual densities (Weigle, Meselson, and Paigen, 1959).

Many rearrangements, including all inversions, will produce a localized reversal of the polarity of the chromosome. It will be recalled that the characteristic direction of synthesis of both mRNA and double-stranded DNA is by addition at the 3' end. In order to maintain the correct polarity, therefore, an inversion must also include an on-axis rotation of 180° (Fig. 8-6). As a result, the direction of transcription of mRNA is reversed in the inverted segment, compared to the rest of the chromosome. In DNA viruses, sometimes only one viral strand is used for transcription: for instance, in coliphage ΦX174 (Hayashi, Hayashi, and Spiegelman, 1964), in *B. subtilis* bacteriophage SP8 (Marmur and Greenspan, 1963), and in *B. megatherium* bacteriophage (Tocchini-Valentini et al., 1963). In other bacteriophages, however, both DNA strands are transcribed

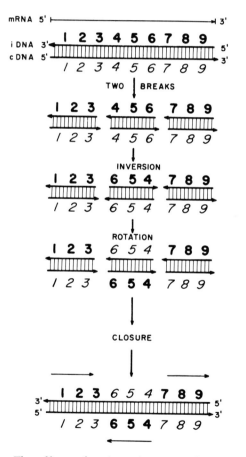

FIGURE 8-6.   *The effects of an inversion upon chromosome polarity and the transcription of messenger RNA, assuming that broken DNA molecules can only rejoin in ways which perpetuate 5' → 3' chemical bonding. The mRNA is transcribed from the iDNA (informational DNA) strand; cDNA refers to the complementary strand.*

(but in different regions of the chromosome), indicating that reversals of the direction of translation have occured. In λ, the left and right halves of the chromosome show opposite polarity (Hogness et al., 1966; Eisen et al., 1966; Cohen and Hurwitz, 1967). In T4, anticlockwise polarity is apparent in the rII genes (Crick et al., 1961; Champe and Benzer, 1962a; McClain and Champe, 1967) and in gene e (lysozyme) (Streisinger et al., 1967), whereas clockwise polarity is apparent in genes 27-51 and 34-35 (Stahl et al., 1966; Nakata and Stahl, 1967) and in the head protein gene (gene 23) (Sarabhai et al., 1964). It is easily seen from Figure 4-3 that a region from about 8 o'clock to about 1-3 o'clock has opposite polarity to the rest of the chromosome. It is therefore possible to speculate that T4 is related by an inversion to some distant ancestor, or possibly even to some extant relative.

9. The origin and properties of macrolesions

Very little is known about the mechanisms which generate large chromosomal deletions, additions, and rearrangements, and what is known is often of a negative nature.

DELETIONS

Deletions occur spontaneously in both bacteria and bacteriophages. About 12 percent of spontaneous T4*r*II mutations are deletions (Benzer, 1959; Folsome, 1962). Their artificial induction in viruses has been reported only once: nitrous acid produces about 8 percent as many deletions in the *r*II region of T4 as it does point mutations (Tessman, 1962*b*). (Beckwith, Signer, and Epstein [1966] have reported that nitrous acid also induces deletions in *E. coli.*) Many other chemical mutagens, on the other hand, fail to induce appreciable frequencies of T4*r*II deletions. Deletions are sometimes induced by ultraviolet irradiation in bacteria (Demerec, 1960), but not in the T4*r*II region (Drake, 1963*b*, 1966*c*).

When many deletions have been mapped in a restricted region, it sometimes becomes apparent that their ends are not distributed at random. The *cys* region in *Salmonella* is particularly susceptible to deletions. Itikawa and Demerec (1967) examined 53 of these deletions, all of whose right ends fell somewhere outside of the *cys* region. All of their left ends fell within a segment which comprised only 8 percent of the *cys* region, but the ends varied slightly in location. Tessman (1962*b*) had previously observed that the left ends of the deletions induced in the T4*r*II region by nitrous acid were located at a small number of preferred locations; their right ends were located outside of the *r*II region. Bautz and Bautz (1967) extended these results by showing that the right ends of these deletions also appeared to be located at a small number of preferred locations. However, while the conclusions of Tessman were based upon mapping studies, those of Bautz and Bautz were based upon measurements of the amount of region-specific messenger RNA which was deleted, and the additional assumption was required that the rate of transcription of DNA was constant in the entire region. In contrast to the induced deletions, spontaneous *r*II deletions do not show a markedly biased localization of ends, and they are also smaller than the induced deletions. Deletions in other microorganisms also frequently possess randomly located ends. It is therefore possible that deletions arise by two (or more) different mechanisms.

Three of the T4rII deletions exhibit an unusually complex composition. These mutants (r196, r1236, and rNB7182) all contain an extended deletion, plus a very closely linked point mutation at one end (Barnett et al., 1967). The point mutation is of the frameshift variety. Terminal "stutters" of this type could actually be very common, but very detailed mapping studies are required to detect them.

A number of different mechanisms can be imagined to generate deletions, and other large lesions as well. These mechanisms involve errors in DNA replication, in genetic recombination, and in DNA repair.

Errors in DNA replication (copy errors) are schematically the most simple (Fig. 9-1). A disruption in the binding between parental template, polymerase,

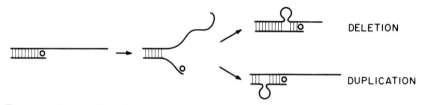

FIGURE 9-1.   *Macrolesions produced by hypothetical errors in DNA replication. Only half of the replication fork is illustrated. The DNA polymerase (circle) detaches from the parental DNA strand; a macrolesion forms if it reattaches at a later point (deletion) or at a previously replicated point (duplication).*

and daughter template could lead to either of two results: a jump ahead in copying, producing a deletion in the daughter strand, or a jump behind, producing a duplication. The template in this case may be taken to be either the entire chromosome, or else only a single strand of DNA. Thus the mutant chromosome may arise either as a heteroduplex heterozygote, or else a homozygote. Copy errors do not necessarily require intrachromosomal homologies, as do the other types of error.

Errors in genetic recombination depend upon intrachromosomal homologies which are sufficiently strong to permit occasional aberrant exchanges (Fig. 9-2). If the homologous regions have the same polarity, then an intrachromosomal exchange will eliminate a small circular element from the chromosome, much as postulated by Campbell (1962) for the escape of episomes from the chromosome. An interchromosomal exchange, on the other hand, will simultaneously produce both a deletion and a duplication. If the homologous regions have opposite polarities, an intrachromosomal exchange will produce an inversion, while an interchromosomal exchange will produce a fairly complicated pair of defective chromosomes. In all of these cases, the mutants will be internally homozygous.

Errors in repair also depend upon intrachromosomal homologies, as well as

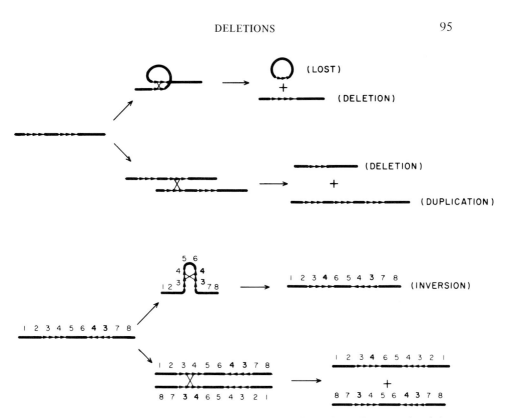

FIGURE 9-2. *Macrolesions produced by hypothetical mechanisms involving recombination. The polarities of homologous segments (indicated by intra-chromosomal arrows) must be parallel in the region of the anomalous recombination event. The line elements in the figure represent complete chromosomes. (The reader may wish to work out the requirements which are imposed on these exchanges when the chromosomes are considered to be double-stranded DNA molecules.)*

upon DNA strand destruction (Fig. 9-3). The likely sequence of events closely resembles a contemporary explanation for the production of frameshift mutations (Chapter 12). A single strand is first digested away in a limited region of the DNA double helix, possibly as a result of base damage and the initiation of a normal repair process. Local strand separation then occurs at one end of the exposed region. Reannealing occurs next, so as to join complementary strands from the two different but homologous regions. Finally, the interrupted strand is covalently closed. The net result will be a heteroduplex heterozygote, the wild type strand being looped out opposite to the strand carrying the deletion. If the digestion step is omitted from this process, and mispairing occurs in the opposite direction, then DNA repair synthesis will produce a duplication heteroduplex heterozygote.

Copy errors have been considered as possible sources of deletions induced by nitrous acid (Tessman, 1962*b*), based on the supposition that either the

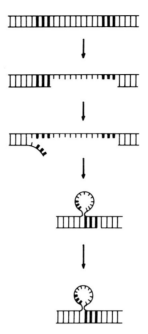

FIGURE 9-3.   *Formation of a deletion by a hypothetical misrepair process.*
(The deletion resides in the bottom half of the final molecule.) A similar scheme
can be constructed for generating a duplication (compare with Figure 12-8).

conversion of guanine to xanthine, or else the cross-linking of complementary
chains, might block the progress of the DNA polymerase; dXTP is not acceptable
in vitro as a substrate (Bessman et al., 1958), and nitrous acid produces
appreciable numbers of linkages between complementary chains (Geiduschek,
1961). However, it is not immediately apparent, without making further
assumptions, why deletions produced by copy errors should exhibit preferentially located ends.

The recombinational models have been assessed in bacterial systems by
comparing deletion frequencies in normal cells and in recombination-deficient
mutants: no differences were detected (Franklin, 1967; Inselburg, 1967). In
bacteriophage T4, where recombination-deficient mutants are not yet available,
it is nevertheless possible to *increase* recombination frequencies with ultraviolet
irradiation, without, however, detectably increasing the frequency of deletions
in the *r*II region (Drake, 1963*b*, 1966*c*). An earlier report had noted an apparent
induction of deletions by ultraviolet irradiation (Folsome, 1962), but the
significance of this observation is not yet clear (see Chapter 14). Although
technically rather difficult, it is in principle possible to determine whether
deletions arise in bacteriophage T4 as a result of interchromosomal exchanges,
that is, by examining deletions which appear among the progeny of a cross
performed with appropriate outside markers.

Barnett et al. (1967) have pointed out that the "stutter" effect can be understood on the assumption that deletions arise between regions of homology (presumably by either the recombinational or the repair route), when the homology is imperfect:

$$\ldots\ldots1.2.3.4.5.6.7.8\ldots\ldots\ldots3.4.5.5.6.9.0\ldots\ldots \rightarrow$$

$$\ldots\ldots1.2.3.4.5.5.6.9.0\ldots\ldots$$

Animal chromosomes contain large numbers of identical sequences, approximately 200 base pairs long, which are distributed throughout the genome (Britten and Kohne, 1968). Although the function of these sequences is unknown, they may mediate the formation of macrolesions of all types by promoting illegitimate crossing over.

In the preceding chapter, a special test was described which makes use of the T4 deletion r1589 to distinguish between nonsense and missense mutations. This test is also able to distinguish between deletions which remove $3n$ versus $3n \pm 1$ base pairs: only in the former case will the activity of the B fragment be preserved. If random numbers of bases are removed, two-thirds of all deletions are expected to be nonsense mutants, since they will induce frameshifts. However, Crick et al. (1961) observed that only three of ten deletions were of the nonsense variety, suggesting that special constraints may exist on the number of base pairs removed. Since then, an additional eight deletions have been tested with opposite results: five of the eight were of the nonsense variety (Table 9-1). The question whether deletions preferentially remove $3n$ base pairs, therefore, remains open.

TABLE 9-1

BEHAVIOR OF DELETIONS IN THE r1589 TEST

Mutant	B Cistron Activity	Mutant	B Cistron Activity
168	−	432	+
184	+	641	−
221	−	680	−
250	+	782	+
386	−	1191	−
1364	+	B45	+
1368	+	H191	−
C33	+	H211	−
EM66	+		
PT153	+		

Presence of B cistron activity indicates deletion of exactly $3n$ base pairs (n an integer) (Data of Drake and Garcia-Molinar). Mutants on the left were described by Benzer (1961), and are presumably the same ten mutants tested by Crick et al. (1961). Mutants on the right were previously described by Benzer (1959).

While the $r1589$ deletion clearly removes an integral number of codons (disregarding uncharacterized elements of intercistronic punctuation), three other deletions which also remove the A/B cistronic division fail to express the function of the B cistron. All three extend farther into the B cistron than does $r1589$, so that it is not immediately obvious whether they delete too much of the cistron, or whether they delete a nonintegral number of codons. However, the B function can be activated in one of these deletions ($r1231$) by inducing frameshift mutations very near to its A-cistron end (Drake, 1963c). The $r1231$ deletion removes about 19 percent of the B cistron, whereas the next-longest deletion, rNB7006, which could not be mutationally activated, removes about 32 percent of the cistron. Thus the dispensable portion of the cistron could be estimated by purely genetic means, in a manner analogous to the determination of the dispensable portion of an enzyme by the progressive removal of amino acid residues from its C- and N-terminal ends.

DUPLICATIONS AND REARRANGEMENTS

Mutations containing duplications are even less well characterized than are mutations containing deletions. Weil, Terzaghi, and Crasemann (1965) have described a class of "partial diploids" of bacteriophage T4 which appear to contain duplications of the rII region. These mutants arise as rare "recombinants" between rII deletion mutants whose lesions *overlap*, so that ordinary recombinants cannot form. (They are produced, for example, in crosses between $r1589$ and $r196$; see Figure 4-5, and the section in Chapter 8 on the $r1589$ test.) In at least one case, the duplication is located near the rII region itself, since the partial diploid condition is carried entirely on a single fragment of DNA (E. K. F. Bautz, personal communication). The partial diploids segregate the parental markers at high rates (up to 63 percent per single growth cycle) with kinetics which resemble the production of recombinants in ordinary crosses. Further studies strongly suggest that the partial diploids often contain an additional deletion in a functionally unrelated region (such as the e gene) (Weil, personal communication). This result is to be expected on the hypothesis that bacteriophage T4 maintains a constant physical chromosomal length: when the duplication exceeds the length of the terminal redundancy, a compensating deletion must be present if viablility is to be preserved. Furthermore, the haploid r segregants should have unusually long terminal redundancies. This prediction has been confirmed (Weil, personal communication): terminal redundancy heterozygotes (specifically, complementation heterozygotes, Stahl et al., 1965) are produced about five times more frequently in crosses between the haploid r segregants than in crosses between the original r mutants.

Mutants which revert at high frequencies, or which produce "revertants" which then throw off the parental mutant at a high frequency, are often suspected of harboring duplications. Freedman and Brenner (quoted in Barnett et al., 1967) have claimed that unstable revertants of certain T4rII frameshift mutants arise by duplications. In addition, some rapidly reverting rII mutants have been shown by Folsome (1964) to appear very early in intracellular virus development, even earlier than do interchromosomal recombinants. (Ordinary mutations tend to appear much later, forming smaller clones; see Chapter 5.) Although he suspected that the mutants contain extraordinarily persistent single-stranded loops, it is more economical to suppose that they contain simple duplications, and that revertants arise by intrachromosomal exchanges. An unstable suppressor mutation in E. coli has been shown to behave like a tandem duplication (Hill et al., 1969). Several extreme-polar mutations (Chapter 18) have been shown to consist of insertions of many hundreds of base pairs; they markedly increase the densities of the bacteriophages in which they reside (Jordan et al., 1968; Shapiro, 1969).

No information is available concerning the mechanism of formation of duplications, but the schemes illustrated in Figures 9-1, 9-2, and 9-3 are all likely candidates.

Since inversions and translocations are not detected sufficiently frequently in viral or other microbial systems to permit an analysis of their origins, only rather special examples of chromosomal rearrangements will be considered here. The specialized transducing bacteriophages have lost a fraction of their genome, and are usually defective as a result. They have also acquired a portion of the host genome, and are thus capable of transduction. Campbell (1962) proposed that this type of particle arises by a recombinational event similar to that depicted in Figure 9-2, the exchange occurring intrachromosomally between homologous regions with the same polarity. However, whereas a prophage sequence 1 . 2 . 3 . 4 . 5 . 6 . 7 . 8 was initially inserted, a sequence W. X. Y. Z. 1 . 2 . 3 . 4 is recovered, where W, X, Y, and Z are host genes. Then, by means of a normal and specific circular permutation, the sequence 3 . 4 . W. X. Y. Z. 1 . 2 is produced. It is consistently observed that the incorporated host genes come from the regions immediately adjacent to the prophage site, but evidence specifically supporting the recombinational hypothesis is yet to be adduced. One important difficulty raised by the Campbell model is to explain how the defective virus is able to survive the original cycle of lytic growth which it initiates. Since its replication normally requires coinfection with a nondefective particle, defective particles may only be recovered when they arise in multinucleate cells.

A rather similar situation is observed with bacterial sex factors (Beckwith, Signer, and Epstein, 1966; Scaife, 1966; Berg and Curtiss, 1967). A single intrachromosomal exchange either removes an integrated sex factor from the bacterial chromosome, or else produces an inversion. When a sex factor recombines out of a bacterial chromosome, it occasionally carries along a portion of that chromosome, leaving behind a deletion. When it recombines into a bacterial chromosome, it may do so with either polarity (clockwise or counterclockwise in the case of the circular E. coli chromosome).

10. Transitions

Hard upon the invention of the duplex model of DNA came the initial understanding of its significance for mutational processes. In a very general sense, mutations would constitute various base pair changes. In particular, the certain infrequent tautomeric forms of the bases were recognized as appealing candidates for the intermediates in the production of transitions (Watson and Crick, 1953*a, b*). In this chapter, evidence for the occurrence and properties of transitions will be considered. Although chemical mutagenesis will be discussed in a later chapter, a few chemical mutagens of very high specificity will also be considered here, since their properties are very closely related to the properties of transition mutations.

In this and future chapters, frequent reference will be made to amino acid, codon, and base pair substitutions. The amino acids and the codons will be abbreviated as in Figure 5-6, the codons being written in the form of ribonucleotide triplets such as UCA = ser (serine). The normal base pairs will be written as A:T or G:C, even when referring to the DNA of the T-even bacteriophages, where cytosine is replaced by glucosylated 5-hydroxymethyl-cytosine. Base pair substitutions will be expressed either in terms of codons (UCA → UUA) or in terms of base pairs (G:C → A:T), depending upon the context of the argument.

MISPAIRING SCHEMES

All four of the common DNA bases can exist in tautomeric forms which are related to each other by single proton shifts. The biologically interesting tautomers involve keto-enol pairs for thymine and guanine, and amino-imino pairs for cytosine and adenine. The infrequent tautomer in each case is capable of regular hydrogen bonding with a normally noncomplementary base, but the pairing is always between a purine and a pyrimidine. (Purine:purine and pyrimidine:pyrimidine mispairings, which will produce transversions, will be considered in the next chapter.) The four resulting illegitimate base pairs are shown in Fig. 10-1, and should be compared with the normal base pairs shown in Fig. 1-2. Genetically, each of the structures in Fig. 10-1 corresponds to a

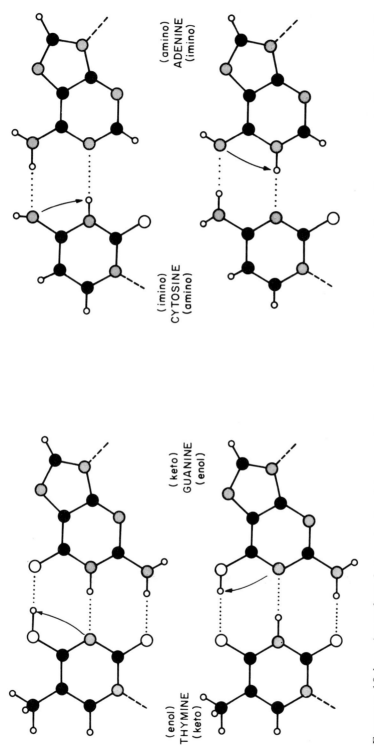

FIGURE 10-1.   *Anomalous base pairs involving rare tautomeric forms of the bases; symbols are the same as in Figure 1-2. Left:
thymine-guanine pairs. Right: cytosine-adenine pairs. The amino and keto forms of the bases are the more frequent tautomers.*

heteroduplex heterozygote, and will segregate in the next DNA replication to produce both a wild type duplex and a homozygous mutant duplex.

Ionization is also capable of generating plausible purine:pyrimidine mispairings (Lawley and Brookes, 1961, 1962). Loss of the proton from the 1-position of either guanine or thymine will allow these two bases to pair through the formation of two hydrogen bonds (Fig. 10-2). Ionized species of cytosine or adenine, however, do not suggest plausible mispairing schemes.

It is instructive to compare the frequencies of the potentially mutagenic tautomers and ionic species with the actual frequencies of transition mutations. Spectrophotometric measurements (Katritzky and Waring, 1962) suggest that the enol form of uracil (and hence approximately of thymine) occurs with a frequency of about 1×10^{-4}. Quantum mechanical calculations (Pullman and Pullman, 1963; Koch and Miller, 1965) suggest that the enol form of uracil occurs with a frequency of about 2.5×10^{-5}, and that the imino form of cytosine occurs with a frequency as high as 10^{-1}. The pK_a of thymine is 9.9 (Shugar and Fox, 1952), and of guanine is 9.2 (Taylor, 1948). At neutral pH,

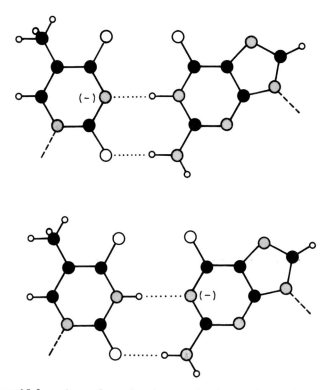

FIGURE 10-2.   *Anomalous thymine-guanine base pairs involving rare ionized forms of the bases; symbols are the same as in Figure 1-2. Protons have been lost from thymine (upper left) and from guanine (lower right) at the negatively charged nitrogen positions.*

therefore, the ionized species should be present at frequencies of about 1.3×10^{-3} and 6.3×10^{-3}, respectively. It will be recalled from Chapter 5 that the mutation rate in bacteriophage T4 is about 2×10^{-5} per gene per replication, that the fraction of mutants containing transitions is less than or equal to 14 percent, and that the overall probability of detecting a base pair substitution is about 3.5 percent in the rII region. Since the average gene contains about 10^3 base pairs, the transition frequency per base pair is about 10^{-7} in T4; it is probably about 10^{-9} in $E.$ $coli.$ Despite the rather approximate nature of both the chemical and the genetic estimates, a gulf several decades deep clearly separates the average probability of mispairing a purine and a pyrimidine and the average probability of generating a mutation. The difference must be explained by assigning to the DNA polymerase an important role in the specificity of base pairing, or else by invoking the action of a rather omniscient repair process. This problem will be discussed further in Chapter 15.

BASE ANALOGUE MUTAGENESIS

A few years after the promulgation of the DNA duplex model, Litman and Pardee (1956) observed that a number of thymine analogues which were fairly efficiently incorporated into the DNA of bacteriophage T4, were also very powerful mutagens. The analogues were halogenated uricil derivatives: 5-bromouracil, 5-chlorouracil, and 5-iodouracil.

Soon thereafter, Freese undertook the study of 5-bromouracil mutagenesis and made two crucial observations. First, all of the mutants induced by 5-bromouracil are in turn revertible by 5-bromouracil or 2-aminopurine, whereas few spontaneous mutants and no proflavin-induced mutants are revertible by base analogues (Freese, 1959a, c). Second, the mutants induced by 5-bromouracil map at very different sites from those occupied by spontaneous mutants, as judged from mutational spectra (Benzer and Freese, 1958; Freese, 1959a; Benzer, 1961a). These observations prompted Freese to define transitions and transversions, adapting the language of quantum mechanics to mutational processes. The mutants induced by 5-bromouracil were assumed to be transitions, presumably on the grounds that the most likely effect of the analogue upon DNA replication would be to induce mispairing with guanine.

Evidence will be presented that 5-bromouracil does indeed induce transitions and not transversions or frameshift mutations. However, the mechanism of mispairing remains unknown. Soon after the original discovery of 5-bromouracil mutagenesis, Meselson (personal communication) pointed out that the electrophilic nature of the bromine atom should increase the frequency of the enol tautomer compared with thymine. Ionization too should be much more frequent: the pK_a of 5-bromouracil is 8.05 (Berens and Shugar, 1963), so that the anion frequency would be 6.6 percent at neutral pH. Faced with such a surfeit of potential mutational events, it is impossible to distinguish the true culprit. However, it is worth noting the results of an in $vitro$ study of mispairing induced by 5-bromouracil (Trautner, Swartz, and Kornberg, 1962): although the reaction was conducted at pH 8.7, where 5-bromouracil should be predominantly in the anionic state, the error rate did not exceed 5×10^{-4}. These results reinforce the conclusion that the DNA polymerase must play a central role in maintaining the specificity of DNA replication.

Other examples exist to illustrate how poorly understood are the details of 5-bromouracil mutagenesis. Fermi and Stent (1962), for instance, reported that the efficiency of mutagenesis of bacteriophage T4 was much higher in singly infected than in multiply infected cells. This difference, furthermore, was not simply a result of the slightly greater number of DNA replications which would occur in singly infected cells, but instead appeared to represent a 2- to 3-fold increase in the probability of mutation per DNA replication. The same authors also observed that the relative rates of mutagenesis at three loci (rII, h, and q) were altered by transient chloramphenicol treatments (300 μg/ml); chloramphenicol blocks most protein synthesis, but allows extended synthesis of biologically active DNA. Intriguing explanations for these results can be constructed in terms of hypothetical properties of DNA polymerase, but the appropriate analyses are yet to be performed.

The appealing simplicity of the mutagenic schemes originally considered for 5-bromouracil prompted Freese (1959b) to examine the mutagenicity of many other base analogues. Only 2-aminopurine and 2,6-diaminopurine were discovered to be mutagenic, however. Mispairing schemes by which 2-aminopurine could induce transitions are illustrated in Fig. 10-3; ionization schemes are not feasible. Two hydrogen bonds can form between 2-aminopurine and thymine (top), and if the 2-aminopurine assumes the infrequent imino form, it can also make two hydrogen bonds with cytosine (middle). However, even the common form of 2-aminopurine can form a single hydrogen bond with cytosine (bottom). The incorporation of 2-aminopurine into DNA occurs at a very low frequency (Wacker, Kirschfeld, and Trager, 1960; Gottschling and Freese, 1961; Rudner, 1961), but the net mutagenicity is at least as strong as is that produced by the much more easily incorporated 5-bromouracil analogue (Freese, 1959b; Champe and Benzer, 1962; Drake, 1963b). Therefore, either the rare tautomeric form of 2-aminopurine is, in fact, relatively much more frequent than is the rare tautomeric form of 5-bromouracil (as is suggested by the quantum mechanical calculations of Pullman and Pullman [1962]), or else the common amino form of 2-aminopurine is the mutagenic intermediate. The difficulty with the second possibility is that it predicts that thymine:guanine base pairs should also form, again mediated by a single hydrogen bond; however, the frequency of this mispairing might be sharply reduced because of the opposing protons at the adjoining nitrogen atoms.

The putative pairing patterns of 2.6-diaminopurine resemble those for 2-aminopurine shown in Figure 10-3, except that a third hydrogen bond could form with thymine, and imino-mediated pairing with cytosine could occur as a result of proton shifts from either the 2-amino or the 6-amino group.

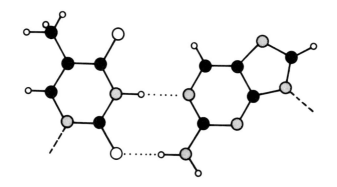

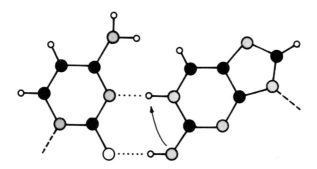

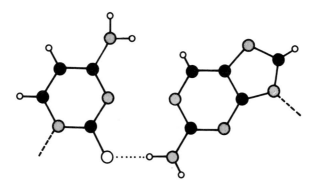

FIGURE 10-3.   *2-Aminopurine base pairing; symbols are the same as in Figure 1-2. Top: pairing with thymine. Middle and bottom: pairings with cytosine.*

The hypothesis that both the halogenated pyrimidines and the aminopurine analogues specifically induce transitions predicts that their mutational spectra should be similar, and this is to some extent true (Benzer, 1961). Spectra were determined for 145 mutants induced by either 2-aminopurine or 2,6-diaminopurine, and for 156 mutants induced by 5-bromouracil. Of the 21 sites occupied by 5-bromouracil mutants, 57 percent were also occupied by aminopurine mutants. However, of the 56 sites occupied by aminopurine mutants, only 27 percent were also occupied by 5-bromouracil mutants. The second correlation is in part weakened because mutants induced by 5-bromouracil show much stronger hot spotting than do mutants induced by the aminopurines; if many more 5-bromouracil mutants were to be mapped, and more sites thereby to be revealed, a greater degree of coincidence with aminopurine sites would probably be revealed. Another factor which undoubtedly tends to separate the two spectra is the directional specificity of the two mutagens. G:C → A:T transitions (identified by hydroxylamine mutagenesis; see next section) are usually induced much more strongly by 5-bromouracil than by 2-aminopurine (in the T4 *r*II system), whereas A:T → G:C transitions are induced much more strongly by 2-aminopurine than by 5-bromouracil (Bautz and Freese, 1960; Champe and Benzer, 1962*b*; Drake, 1963). These specificities presumably reflect in part the different tautomerization or ionization probabilities of the analogues in polynucleotides and in the free nucleoside triphosphate state. In the S13–ØX174 system, on the other hand (Howard and Tessman, 1964*a*), base analogue specificities seem to be just the opposite: A:T → G:C is preferentially induced by bromouracil, and G:C → A:T by aminopurine. Other factors, such as the nature of the DNA polymerase itself, probably influence the mutagenic specificities of base analogues.

HYDROXYLAMINE MUTAGENESIS

Before proceeding to a consideration of the evidence that base analogues induce only transitions, and not transversions or frameshift mutations, it will be useful to consider the properties of the most highly specific mutagen thus far described. Since base analogues are capable of reverting the same mutants which they induce, they presumably generate transitions in both directions. A search was therefore initiated by Freese for chemical mutagens which might act with great specificity upon nonreplicating DNA, in the sense that only a single base would be mutated. Hydroxylamine (NH_2OH) was discovered to be a powerful mutagen (Freese, Bautz-Freese, and Bautz, 1961) with the required specificity.

Hydroxylamine attacks both cytosine and uracil (in RNA) and, to a much lesser extent, thymine (Freese, Bautz, and Freese, 1961; Freese, Bautz-Freese, and Freese, 1961; Brown and Schell, 1961; Schuster, 1961). However, two lines of evidence strongly indicate that only the reaction with cytosine is mutagenic. First, the reaction with cytosine is favored at low pH, whereas the reaction with uracil (and hence presumably with thymine) is favored at high pH (Schuster,

1961). Concomitantly, mutagenesis is favored at low pH (Champe and Benzer, 1962*b*; Schuster and Wittmann, 1963). Thus at pH values of 6, 7.5, and 9, the relative reaction rates with cytosine were 32:13:<4, whereas the relative reaction rates with uracil were <1:13:30; the relative mutation rates measured by the reversion of T4*r*II mutants were 28:12:1. Second, while T4*r*II mutants induced with hydroxylamine are uniformly induced to revert with base analogues, they are not induced to revert with hydroxylamine itself (Freese, Bautz, and Freese, 1961; Champe and Benzer, 1962*b*). Furthermore, a close study of mutants of bacteriophages S13 and ΦX174 induced by a variety of agents in both the forward and reverse directions, revealed that hydroxylamine exhibited the greatest specificity of any of the agents employed (Tessman, Poddar, and Kumar, 1964; Howard and Tessman, 1964*a*). One important advantage of this S13–ΦX174 system derives from the single-stranded DNA of the virus particles: guanine and cytosine occur separately, instead of being inevitably paired. Host range mutants induced in these viruses by hydroxylamine were not induced to revert by hydroxylamine, and a comparison of mutations induced by hydroxylamine with mutations induced by nitrous acid, ethyl-methanesulfonate, and base analogues strongly implied that only cytosine was mutated by hydroxylamine.

More recent evidence confirming the mutational specificity of hydroxylamine has come from studies of the chain-terminating codons, UAA (ochre), UAG (amber), and UGA. All three of these codons can be generated from the wild type by hydroxylamine, but none are induced to revert with hydroxylamine (Champe and Benzer, 1962*b*; Brenner, Stretton, and Kaplan, 1965; Brenner et al., 1967). The lack of reversion reflects the fact that the codons which do contain guanine (and hence contain cytosine in the corresponding DNA position), are connected at those positions by transitions to the ochre triplet. By the use of specific suppressor host cells, it was also shown that the transition UAA → UAG was induced by 2-aminopurine but *not* by hydroxylamine, whereas the transition UGA → UAA was induced by both agents. Therefore, not only does hydroxylamine *not* mutate adenine or thymine, but it also mutates cytosine *only* by inducing transitions.

Since hydroxylamine is capable of reacting with a wide variety of molecules within the cell compared to its reactivity with DNA, its specificity might be expected to disappear when tested *in vivo*. Tessman, Ishiwa, and Kumar (1965) reported that transitions were indeed initiated in all possible directions in the S13–ΦX174 system when infected cells were treated; they also observed that pretreatment of the cells (before viral infection) was sufficient to achieve *in vivo* mutagenesis.

AMINO ACID SUBSTITUTIONS

The codon catalogue (Figure 5-6) is now sufficiently well established to provide a basis for interpreting the base pair changes which underlie amino acid substitutions. We will therefore consider the amino acid substitutions induced by base analogues. Although it is implicit in what follows, it should be explicitly noted that the great majority of amino acid substitutions which have been detected in a variety of systems during the preceding decade, are explicable on the basis of single base pair substitutions. The amino acid substitutions themselves also occur singly, except in the regions between a pair of mutually suppressing frameshift mutations (see Chapter 12). Mutational mechanisms involving multiple base pair alterations in a single elementary act need not therefore be considered.

Despite the extensive use of base analogue mutagenesis to obtain mutants with amino acid substitutions, both the mutagen and the resulting amino acid substitution have been specified in discouragingly few instances. The most extensive available data derive from studies on the tryptophan synthetase A protein of *E. coli*, and have been summarized by Yanofsky, Ito, and Horn (1966). These results are summarized in Fig. 10-4 in the form of mutational pathways written in the language of amino acids and their mRNA codons. It is clear from the figure that 2-aminopurine is capable of inducing the transition $A:T \rightarrow G:C$, but no information is provided about the reverse transition. More important, however, is the conclusion that none of the four possible transversions ($G:C \rightarrow T:A$, $G:C \rightarrow C:G$, $A:T \rightarrow C:G$, $A:T \rightarrow T:A$) are appreciably induced; the infrequent transversions (3/32) which did appear among 2-aminopurine-induced revertants of mutants A23 and A46 are readily attributable to contamination by the transversion-rich spontaneous background.

Reversion patterns from the amber (UAG) codon also confirm the ability of 2-aminopurine to induce the transition $A:T \rightarrow G:C$ (Stretton, Kaplan, and Brenner, 1966). Amber mutations in the structural gene for bacteriophage T4 head protein were induced to revert both to glutamine (UAG → CAG) and to tryptophan (UAG → UGG). Whereas all 21 of a series of spontaneous revertants of *am*H36 consisted of transversions to tyrosine (UAG → UAC or UAU; $G:C \rightarrow T:A$ or C:G), 9 out of a series of 10 2-aminopurine-induced revertants contained glutamine; the tenth contained tyrosine, and probably arose from the spontaneous background. It is again clear that 2-aminopurine induces transversions either very poorly, or else not at all.

GENETIC TESTS OF MISPAIRING HYPOTHESES

A number of experiments have been attempted to distinguish between errors of replication (directed by base analogue residues already located in the polynucleotide) and errors of incorporation (directed by free base analogues). The specific mispairing schemes to be considered appear in Figs. 10-5 (for 5-bromouracil) and 10-6 (for 2-aminopurine), where it may be seen that both base analogues induce errors of replication along the $A:T \rightarrow G:C$ pathway, and errors of incorporation along the $G:C \rightarrow A:T$ pathway. These schemes lead to two types of predictions.

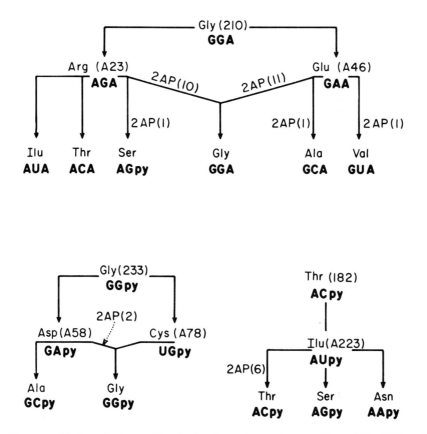

FIGURE 10-4.   *Amino acid substitutions observed at positions 182, 210, and 233 in the E. coli tryptophan synthetase A protein. 2AP = substitutions induced by 2-aminopurine, with number of occurrences noted in parentheses; py = U or C; A23 = mutant designation. The bottom line of each array lists amino acids which produce a partially or fully active protein.*

First, a transient exposure of replicating DNA to base analogues should distinguish A:T and G:C base pairs according to whether new mutations continue to appear after removal of the mutagen. (Here "removal" implies not only removal of the exogenous source, but also depletion of intracellular pools and avoidance of DNA degradation and reutilization.) A:T → G:C transitions can occur at any DNA replication after the analogue has been incorporated, but G:C → A:T transitions can only occur in the presence of the free base analogue. The two situations are often called "clean-growth" and "dirty-growth" mutagenesis, respectively.

Second, the generation at which induced mutations first begin to appear should depend upon both the base analogue and the target base pair. In order to clarify the counting of generations we will make two simplifying assumptions: the mutagen is available to the target base pair immediately after it is added; and

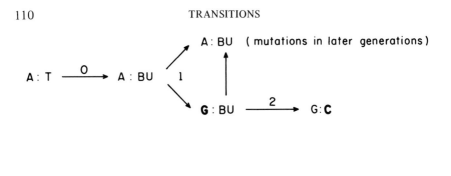

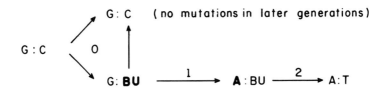

FIGURE 10-5.   *5-Bromouracil mutagenesis. The first opportunity for a base to direct the transcription of mutant messenger RNA is indicated by bold-face symbols. The generation at which this can occur depends upon the purine-pyramidine orientation of the affected base pair.*

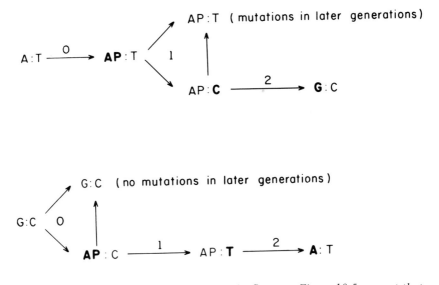

FIGURE 10-6.   *2-Aminopurine mutagenesis. Same as Figure 10-5, except that the first opportunity for the purine-containing strand to direct the transcription of mutant messenger RNA remains in doubt.*

the resulting mutations are expressed immediately after the appearance of a "mutant" base on the DNA strand used for transcription. The number of generations required to produce a "mutant" base for transcription depends upon the direction of the transition, upon the base analogue used, and upon the orientation of the base pair (G:C in this case behaving differently from C:G, and A:T from T:A). Since 5-bromouracil can often completely replace thymine without inducing extensive lethality, it must be transcribed like thymine with high probability. Thus when A:BU misreplicates to G:BU (one generation), mutation expression occurs immediately if guanine is the transcribed base, whereas another generation must ensue if cytosine is to be the transcribed base. When 5-bromouracil misincorporates to form G:BU (zero generations), mutation expression occurs immediately if bromouracil is the transcribed base, whereas another generation must ensue if adenine is to be the transcribed base. The situation is less clearly defined with 2-aminopurine. The analogue can be incorporated into DNA to about 1 percent without extensive lethality. Since this corresponds to about 20 residues per gene, it must be transcribed normally much of the time (presumably as adenine). However, it may occur in the "mutant" form with sufficiently high frequency to allow mutation expression. In Figures 10-5 and 10-6, the first possible appearance of "mutant" bases is indicated by bold face symbols.

The question of when induced mutants should first appear is more simply answered than the question of how long thereafter mutants should continue to appear during clean-growth mutagenesis. A mutation just induced by 5-bromouracil, in the form of a G:BU heterozygote, may be detected with an efficiency of 1.0, but after a single division its frequency should drop to 0.5 as the heterozygote is resolved into homozygous mutant and wild type. If the mutant arises in a multi-nuclear cell, nuclear segregation will also decrease the apparent mutant frequency. Dilution of early-appearing mutants by segregation may be just sufficient to offset the continued appearance of new mutants.

The first evidence for the existence of clean-growth mutagenesis was reported by Litman and Pardee (1960), who observed that forward mutations induced in bacteriophage T2 by 5-bromouracil continued to appear after base analogue incorporation was relieved by adding thymidine. The data of Rudner (1960, 1961) on the induction of reversions of a tryptophan auxotroph of *Salmonella typhimurium* by 2-aminopurine), also strongly suggested the occurrence of clean-growth mutagenesis. However, the first clear separation of dirty-growth and clean-growth responses was obtained by Strelzoff (1961, 1962), who studied the induced reversion of *E. coli* auxotrophs. Revertants of a methionine auxotroph continued to appear for several generations after the removal of the base analogue (either 5-bromouracil or 2-aminopurine), indicating an A:T→C:C pathway. Revertants of five other auxotrophs ceased to appear after the removal of the base analogue, indicating G:C → A:T pathways.

A very striking demonstration of clean-growth mutagenesis was also obtained using *e* (lysozyme) mutants of bacteriophage T4 (Terzaghi, Streisinger, and

Stahl, 1962). A series of mutants was grown in the presence of 5-bromouracil, and the yield of e^+ revertants was taken as a measure of dirty-growth mutagenesis. These particles were then used to infect cells at a low average multiplicity in the absence of 5-bromouracil. Shortly after the appearance of mature intracellular particles, chloroform was added. Cells infected with preexisting (dirty-growth) e^+ revertants promptly lysed, and the liberated particles were washed away; the remaining infected cells did not lyse, even when a (clean-growth) e^+ revertant originated within them. The clean-growth yield was obtained later by lysing the cells with egg white lysozyme; its revertant content was then determined. In this way, two e mutants were observed to produce unequivocal clean-growth responses, while four other mutants showed no clean-growth responses. Unfortunately, the behavior of these six mutants was anomalous in other respects. For instance, one of the mutants exhibiting clean-growth mutagenesis (and therefore supposedly containing A:T at the target site) was induced to revert with hydroxylamine, while the four mutants supposedly containing G:C at the target site were uniformly nonresponsive to hydroxylamine. It would be very helpful to reassess the properties of these mutants, for instance by attempting stronger hydroxylamine treatments, and by investigating the role of suppressor mutations in the reversion process (see Chapter 17).

The earliest studies of the kinetics of appearance of mutants induced by base analogues (Rudner, 1960, 1961a, b) produced equivocal results because of uncertainties in the counting of generations: differently prepared cells were used with different base analogues, and assumptions had to be made about the number of DNA replications which occurred on the assay plates before rigorous selection was established. However, revertants of the $S.$ $typhimurium$ tryptophan auxotroph D-79 first appeared two or three generations after 5-bromouracil treatment, but only one or two generations after 2-aminopurine treatment. Clean-growth mutagenesis was also clearly observed after 2-aminopurine mutagenesis. These results are most fully compatible with an A:T $\rightarrow$ G:C pathway, in which the pyrimidine is the transcribed base; furthermore, clean-growth mutagenesis was observed with 5-bromouracil.

Strelzoff (1962) studied the kinetics of reversion of five $E.$ $coli$ auxotrophs. In four, the revertants appeared one generation after base analogue treatment, whereas in the fifth, revertants first appeared after two generations; the fifth mutant was also the only one to exhibit clean-growth mutation. She therefore concluded that the first four mutants contained G:C base pairs, and the fifth contained an A:T base pair. In contrast to the predictions of Figures 10-5 and 10-6, however, no base analogue-dependent differences in the kinetics of expression were observed with any of the mutants (Strelzoff and Ryan, 1962). This paradox remains unresolved.

An even more complicated and confusing set of results has been reported by Margolin and Mukai (1961, 1966), who studied the reversion of leucine auxotrophs of $Salmonella$ $typhimurium.$ Two major groups of mutants could be distinguished. Group I contained mutants originally induced by 2-aminopurine, and rather weakly reverted by 2-aminopurine. A number of divisions following mutagenesis by 2-aminopurine were required before the induced revertants appeared. Even more divisions produced no further increases in the revertant frequency. It was therefore not clear whether the requirement for limited cell division was in fact a requirement for phenotypic expression (Figure 10-6), or for clean-growth mutagenesis partly concealed by segregation-dilution. The two types of expression kinetics expected for 2-aminopurine mutagenesis could not be discerned among these mutants. Ronen (1964) observed that the Group I mutants were induced to revert by diethylsulfate (see Chapter 13) only after a long delay without cell division (and therefore not necessarily correlated with DNA replication). While Ronen believed that these mutants contained A:T base pairs, Margolin and Mukai suspected that they contained G:C base pairs. The properties (and authors' interpretations) of the Group II mutants were the converse of the first group. These mutants were efficiently induced in the forward direction by 5-bromouracil, and were strongly reverted by 2-amino-purine within one or two divisions. Revertants induced by diethylsulfate also appeared promptly. Unfortunately, not enough hard information exists about these two groups of mutants to deduce with confidence their molecular compositions. A further complication which may have to be considered is the repair of heterozygotes which may occur before chromosomal replication (see Chapter 16). Margolin and Mukai (1966) have attempted to interpret their results by assuming that transcription normally reads both DNA strands simultaneously, information transfer depending upon the portions of the bases which extend into the "narrow groove" of DNA (Figure 1-3). Strelzoff and Ryan (1962) made rather similar speculations. If this interpretation is correct, then the simple schemes summarized in Figures 10-5 and 10-6 are not applicable to the problem of the kinetics of mutation expression.

A rather different test of the mispairing hypothesis was performed by Pratt and Stent (1959), who sought evidence for the existence of the mispaired G:BU intermediate itself. Cells infected with rII mutants of bacteriophages T2 or T4 were exposed to 5-bromouracil, and very shortly thereafter prematurely lysed. The resulting r^+ revertants were primarily (80 percent) r/r^+ heterozygotes, in agreement with their probable G:BU composition. If lysis was delayed, the majority of revertants appeared as homozygotes. The properties of mutational heterozygotes will be discussed in more detail in Chapter 16.

CHEMICAL TESTS OF MISPAIRING HYPOTHESES

Readers who may have concluded that the amino acid substitution data and the various genetic tests of mispairing hypotheses hardly provide a firm experimental foundation for the theory of transitions will find little cheer in the meager chemical data.

In a direct *in vitro* attack upon the mechanism of 5-bromouracil mutagenesis, Trautner, Swartz, and Kornberg (1962) examined the accuracy with which the *E. coli* DNA polymerase synthesized DNA using single-stranded poly-AT (.... ATATAT ...) and poly-ABU templates. The DNA polymerase is capable of converting these templates to the double-stranded form by synthesizing the complementary strand from nucleotide triphosphates. (The *in vivo* role of this enzyme in DNA synthesis remains obscure, since it is only capable of synthesis in the $5' \rightarrow 3'$ direction, whereas the *E. coli* chromosome is replicated simultaneously in both directions [Cairns, 1963]). When poly-AT was used as a template in the presence of ATP, BUTP, and very highly radioactive GTP substrates, less than one guanine was incorporated per 10^5 total residues. The reaction was carried out at a high pH, which might have induced a much higher error rate because of ionization. When poly-ABU was used as a template, however, guanine was incorporated at a frequency of between 4×10^{-5} and 5×10^{-4} per total residues incorporated.

Difficulties arose when the sites of guanine incorporation were examined. When a substrate nucleotide triphosphate is labeled with P^{32} in the α position (adjacent to the sugar), and the product is hydrolyzed with spleen phosphodiesterase, the label is transferred to the preceeding base (Josse, Kaiser, and Kornberg, 1961) (Figure 10-7). On the assumption that guanine mispairs with 5-bromouracil, the preceding base should invariably have been bromouracil. This occurred on the average only 41 percent of the time, however, the remainder of the guanine residues following adenine (17 percent) or guanine itself (42 percent). Among the more obvious explanations for these results might be that the *in vitro* system permits a wider spectrum of mispairings than does the *in vivo* system, or that a significant portion of the *in vitro* synthesis, and especially the error-prone portion, occurs by end-addition to the template strand or by reversal of the enzyme onto its own product (Richardson, Schildkraut, and Kornberg, 1963). The superior enzyme preparations which are now available (Goulian and Kornberg, 1967) justify an experimental reevaluation of the results.

Another approach to the analysis of mispairing involved a combination of genetic and chemical methods (Howard and Tessman, 1964*a*). A number of specific transitions in the S13–ΦX174 system were identified by their induction by base analogues, hydroxylamine, and other mutagens. The induction of A → G and T → C transitions by bromodeoxyuridine was strongly suppressed by exogenous TdR (thymidine deoxyribonucleotide), but not with any regularity by GdR, AdR, or CdR. The induction of the reverse transitions, however, was not repressed by TdR, but was repressed by CdR. While the data presented are extremely variable, the results are consistent with the supposition that G:C → A:T transitions induced by bromouracil depend upon the replacement of cytosine residues, while A:T → G:C transitions depend upon the replacement of thymine residues. When similar experiments were attempted using 2-aminopurine, no consistent repression of the induction of A:T → G:C transitions was observed with any deoxyribonucleotide, but the induction of G:C → A:T

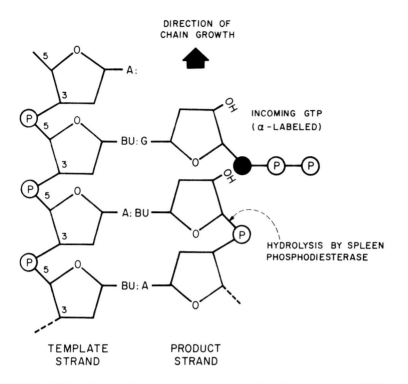

FIGURE 10-7.   Determination of nearest neighbor frequencies in DNA. The P³² atom attached to the incoming GTP residue will be transferred to the preceding bromouracil residue after nuclease digestion.

transitions was moderately repressed by GdR. These results are at least consistent with the supposition that G:C → A:T transitions may be mediated by the replacement of guanine residues by 2-aminopurine.

SUMMARY

The hypothesis that base analogues induce transitions—and only transitions—has achieved great popularity. They clearly *do* induce transitions, and clearly *do not* induce frameshift mutations. The experimental evidence to suggest that they *never* induce transversions, and the evidence to suggest that they actually induce transitions by a simple mispairing process, remains meager.

   A number of areas are clearly overripe for renewed investigation. Much more extensive data are needed on amino acid substitutions induced by 2-amino-purine, and especially by 5-bromouracil. The kinetics of mutation expression in bacterial systems seriously need to be reexamined, using as wide a variety of experimental approaches as possible. Finally, the *in vitro* analysis of mistakes induced by base analogues in DNA synthesis should be reinitiated.

11. Transversions

In the previous chapter, the properties of transitions were examined in conjunction with the properties of a small number of highly specific chemical mutagens. While it has been shown unequivocally that transversions arise spontaneously, and may also be induced by various mutagenic treatments, they nearly always appear along with transitions, and sometimes even with frameshift mutations. Furthermore, as we shall see, much less is understood about the mechanisms which generate transversions than about the mechanisms which may generate transitions.

GENETIC CRITERIA
FOR RECOGNIZING TRANSVERSIONS

Of the three elementary types of mutational microlesions, frameshift mutations are generally induced to revert with acridines and related compounds (see Chapter 12), and transitions are induced to revert with base analogues. Mutants which are induced to revert with neither agent, but which do revert spontaneously, are therefore likely to be transversions. This is at best only a supposition, but it takes on added weight if the mutants can be shown to revert under one or more of the conditions (to be discussed below) which can induce transversions. This "nonresponse" criterion for identifying transversions is supported by two types of evidence. First, mutants identified by their corresponding amino acid substitutions as having arisen by transitions are uniformly capable of being induced by base analogue mutagenisis, whereas amino acid substitutions not inducible by base analogues have uniformly resulted from transversions (Stretton, Kaplan, and Brenner, 1966; Weigert, Gallucci, Lanka, and Garen, 1966; Yanofsky, Ito and Horn, 1966). Second, mutants produced by agents which are believed to produce only transitions (or frameshift mutations) are uniformly induced to revert by base analogues (or by acridines). Thus virtually all T4rII mutants induced by 5-bromouracil, 2-amino-purine, nitrous acid, and hydroxylamine are induced to revert by at least one of the base analogues (Bautz-Freese and Freese, 1961; Champe and Benzer, 1962b; Freese, 1959c; Freese, Bautz-Freese, and Bautz, 1961; Orgel and Brenner, 1961). T4rII mutants induced by acridines, on the other hand, are also induced to revert by acridines but not by base analogues (Orgel and Brenner, 1961; Drake, unpublished data).

Unfortunately, a very serious complication limits the reliability of the "nonresponse" test for the identification of transversions. We have seen earlier that many amino acid substitutions may be relatively inocuous, depending upon the variability which can be tolerated by a particular protein without serious loss of function. Many mutants which arise as transversions probably can revert by transitions at the same site, or else within the same codon. Although an amino acid different from the wild type may result, function will frequently be restored. Therefore, when a collection of mutants induced by a particular mutagen is found to contain a significant proportion of mutants which are *not* reverted by base analogues (or by acridines), then many of the mutants which are induced to revert by base analogues are also likely to contain transversions. It may be very difficult to determine by either genetic or chemical means which of the responding mutants are really transitions, and which are transversions.

In principle it should be possible to establish whether an agent is capable of inducing at least certain transversions by testing its action on selected chain-terminating mutations. Viral ochre mutations induced by base analogues or hydroxylamine arise solely by the CAA → UAA route, the other two positions of the amber codon being related by transitions to the UGA and UAG codons, respectively. If such an ochre mutant can be induced to revert to a form which is phenotypically distinguishable from the true wild type, then the reversion must have occurred along a transversional pathway, and must, in fact, have originated from an A:T base pair. The UGA and UAG codons are each related to two amino acids by transitions, so that it would be necessary to show that insertion of either amino acid (arginine and tryptophan, and glutamine and tryptphan, respectively) produced fully functional revertants before these codons could be used in a similar manner.

TRANSVERSIONS RECOGNIZED
BY AMINO ACID SUBSTITUTIONS

The amino acid substitutions which have been recognized in the *E. coli* tryptophan synthetase A protein system (Yanofsky, Ito, and Horn, 1966) were summarized in Figure 10-4. Most of the observed substitutions were in fact attributable to transversions: A23 (arginine) to serine, threonine and isoleucine; A46 (glutamic acid) to valine and alanine; A58 (asparagine) to alanine; A78 (cysteine) to glycine; and A223 (isoleucine) to serine and asparagine. When translated into the language of the base pairs, these mutations include all of the possible transversions (C:G ↔ G:C ↔ T:A ↔ A:T). Most of these transversions (G:C ↔ T:A ↔ A:T) were observed to occur spontaneously. The alkylating agent ethylmethaesulfonate (EMS) also induced at least some of them (G:C ↔ C:G and A:T ↔ T:A).

An extensive study of revertants recovered from amber and ochre codons in the *E. coli* alkaline phosphatase protein is summarized by Weigert et al. (1966). Both codons can mutate by transversional pathways to five different amino acids (glutamic acid, lysine, serine, leucine, and tyrosine), and all five substitutions were observed in revertants from various sites.

Stretton, Kaplan, and Brenner (1966) studied spontaneous revertants of T4 head protein amber mutatants. A surprisingly large fraction of the revertants contained tyrosine, which corresponds to the G:C → C:G or the G:C → T:A transversion. At eight sites where tyrosine was definitely an acceptable amino acid, 78 percent of the revertants contained tyrosine, and in a few cases all of the revertants (among about 20 tested) contained tyrosine. These data, in contrast to those available from the two *E. coli* systems, suggest a marked bias in transversional specificity.

THE INDUCTION OF TRANSVERSIONS

The first mutants to behave like tranversions in reversion tests were detected in the T4rII system (Freese, 1961). A number of mutants (16 of spontaneous origin, and five induced by ethylethane sulfonate [EES]) which were not induced to revert by base analogues but which reverted spontaneously, were tested for their responses to EES and to hot acid (45°, pH 5, which is certainly hot acid from the standpoint of the virus). None of the spontaneous mutants but two of the EES-induced mutants were induced to revert, always by both agents. Since nine out of ten proflavin-induced mutants failed to respond, it is unlikely that hot acid or EES induces frameshift mutations at an appreciable rate. Freese's conclusion that the observed responses were due to transversions is therefore the most probable one.

It is already clear from studies of T4rII mutants, and from observed amino acid substitutions, that a substantial proportion of spontaneous mutations may be transversions. While most spontaneous mutants arise during the growth of a culture, it should be noted that spontaneous mutations also appear during the storage of nonreplicating genomes (Drake, 1966a). The only T4rII mutants observed to revert during storage were also capable of responding to hydroxylamine, and therefore contain G:C base pairs. However, forward mutations arising in the absence of DNA replication comprised a complicated mixture (Drake and McGuire, 1967a). A very small fraction may have contained frameshift mutations. About two-thirds of the mutants were revertible with base analogues, while about one-third was refractory both to base analogues and to acridines. These presumably contain transversions, and therefore, as argued previously, the mutants revertible by base analogues also frequently contain transversions. A large majority of the mutants responding to base analogues also responded to hydroxylamine, suggesting that one of the most frequent tranversions observed was G:C → C:G. This conclusion is supported by the observation that stocks of T4rII amber (UAG) mutants, but not of ochre (UAA) mutants, accumulate revertants at 44° over a period of a week (Krieg, personal communication).

T4rII mutations can also be readily induced by a combination of white light irradiation and photosensitizing dyes, and closely resemble mutations accumulated spontaneously in the absence of DNA replication (Drake and McGuire, 1967b). About two-thirds were induced to revert by base analogues, and about half of these were also induced to revert with hydroxylamine. The remaining third of the mutants was refractory both to base analogues and to acridines. Although these results strongly suggest that transversions may be induced photodynamically, they do not give any reliable indication of the specific pathways involved.

The most highly specific source of transversions yet recorded is a remarkable *E. coli* mutator mutation (Yanofsky, Cox, and Horn, 1966; Cox and Yanofsky, 1967). Mutator mutations, which will be discussed more fully in Chapter 15, induce increased spontaneous mutation rates throughout the genome. This *E. coli* mutator increases the spontaneous mutation rate about 1000-fold, but a close analysis of associated amino acid substitutions in the tryptophan synthetase A protein indicates that only a single mutational pathway is involved, namely the transversion A:T → C:G. For instance, of the substitutions illustrated in Figure 10-4, the mutator promotes the appearance of the serine revertant from A223, the glycine revertant from A78, the alanine revertant from A46, the serine revertant from A23, and the alanine revertant from A58. Unfortunately, this mutator acts very weakly upon bacteriophage T4 (Pierce, 1966), and not necessarily with the same mutational specificity.

Gene 43 of bacteriophage T4 specifies an enzyme involved in DNA polymerization. Many of the mutations of this gene also produce mutator activity. While several of these mutators produce transversions, none do so both strongly and exclusively (Speyer, Karam, and Lenny, 1966; Allen and Drake, unpublished results).

MECHANISMS FOR THE PRODUCTION OF TRANSVERSIONS

Although a number of mechanisms can be imagined for the production of transversions, one, depurination, has been repeatedly invoked by students of mutation (Freese, 1959c). Although the reasons for this will be apparent, it is an unfortunate situation, since a number of other appealing possibilities have been largely overlooked, and since depurination itself is not at all well established as a transversional mechanism.

Depurination has been observed to result from two quite different treatments to DNA, namely heat, especially at low pH (Greer and Zamenhof, 1962), and alkylation (Bautz and Freese, 1960). Since both treatments are mutagenic, and even appear to effect the same transversions, depurination seemed correlated with mutation. However, it was necessary to imagine that a gap in the sequence of DNA bases could result, during DNA replication, in the somewhat indiscriminate insertion of a base in the complementary strand. It is also very likely that numerous other (as yet uncharacterized) reactions occur both in heated and in alkylated DNA. It is already clear, for example, that heating at low pH oxidatively deaminates cytosine to produce uracil (Shapiro and Klein, 1966), and thus presumably generates G:C → A:T transitions. Alkylation of guanine, in addition to inducing depurination, also enhances ionization (Lawley and Brookes, 1961), and may thus generate the same transition.

The alkylation-induced depurination of guanine is a slow process (Lawley and Brookes, 1963), and is readily separable from the primary reactions themselves (Ronen, 1968). It is therefore feasible to ask whether an organism, once treated with an alkylating agent, accumulates further damages during post-treatment incubation. The data bearing on this question are somewhat contradictory, and are probably made more difficult to interpret because of repair processes which may act upon alkylated DNA (Papirmeister and Davison, 1964; Strauss and Reiter, 1965). In bacteriophages T2 and T4, the full yield of mutants has been reported to appear immediately after EMS or diethylsulfate (DES) treatment; post-treatment incubation for as much as 21 hours at 47° adds numerous lethal hits, but no mutational hits (Strauss, 1961; Lobbecke and Krieg, quoted in Krieg, 1963a; Ronen, personal communication). (These results should not be confused with those of Green and Krieg [1961] and Krieg [1963b], who observed delayed appearance of mutations *during replication* of alkylated T4 DNA, probably as the result of mispairing by alkylated bases.) However, it would be premature to conclude from these results alone that depurination is not mutagenic, since the majority of mutations induced in T4 by alkylation consist of transitions (Bautz and Freese, 1960; Freese, 1961; Krieg, 1963b). Until the experiments are confirmed for a transversional pathway, it will be possible to suppose that a few transversions do arise during post-treatment incubation, but are not observed through the background of transitions.

In contrast to viral systems, post-alkylation incubation of starved stationary-phase bacteria may generate mutational lesions. Strauss (1962a) observed reversion of *E. coli* auxotrophs after treatment with EMS, but not after treatment with methylmethanesulfonate (MMS). Ronen (1964) observed two classes of leucine auxotrophs in *Salmonella*, one exhibiting the full reversional response immediately after treatment with DES, and the other requiring additional incubation. The first class was considered to result from G:C → A:T transitions, and the second class to result from depurination. However, at least two other possibilities are yet to be eliminated: reactions other than depurination (for instance base rearrangements), and imperfect attempts by starved and stationary phase cells to repair alkylated residues.

Mechanisms for generating transversions can be divided into null-base schemes and mispairing schemes. Null-base schemes result from modifications of DNA which either remove a base entirely (such as depurination), or else so extensively degrade it that it no longer resembles any base. If depurination is indeed not mutagenic, then all null-base schemes are probably sterile. Mispairing schemes, on the other hand, have yet to be considered in any detail. They may depend upon two quite different processes: mispairing between normal bases themselves, and mispairing promoted by the chemical modification of a normal base.

Mispairing by normal bases might occur by either of two processes. If a single-strand loop in the parental strand appeared momentarily and reversibly at the replication site, it might prematurely align a base in a position where it could pair normally with the next base to be added to the progeny strand (Kapuler, personal communication); the rapid return of the parental strand to its normal configuration would be necessary to prevent the formation of a deletion mutation (see Chapter 9) or of multiple adjacent base pair substitutions, which are almost unknown. This process, which is compatible with model building when the choice of a new base is performed exclusively by the bases near the growing point, could clearly generate base pair substitutions of all types. Alternatively, direct mispairing might occur between bases at the same level. A large number of hydrogen-bonded configurations are possible between bases in addition to those normally encountered in DNA (Donohue, 1956), and some of these occur normally in RNA, particularly in the codon:anticodon reconition

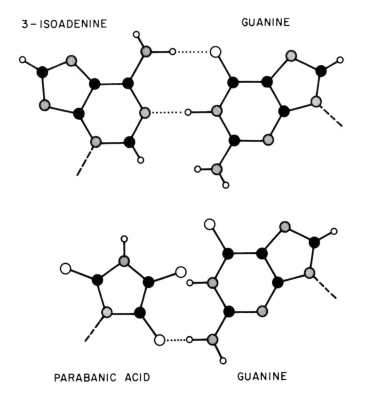

FIGURE 11-1.   *Hypothetical base pairs mediating the formation of transversions; symbols are the same as in Figure 1-2.*

process (Crick, 1966). The fact that these mispairing schemes violate the integrity of the conventional double helical B structure of DNA is not a significant argument against their occurrence: mutation is a rare and abnormal process no matter how it occurs, and certainly involves the formation of improbable intermediates (see Chapter 16).

Conceivably, mispairing schemes may also depend upon the degradation or rearrangement of a purine to form a structure which will pair with another purine, without necessarily violating the dimensions of the double helix. Thus 3-isoadenine, in which the glycosidic bond involves N3 instead of N9, and which is accepted in a variety of enzymatic reactions including polymerization (Leonard and Laursen, 1966; Michelson et al., 1966), can form a pair of hydrogen bonds with guanine to produce an intraduplex bridge of standard length. Parabanic acid, a photochemically produced degradation product of guanine (Sussenbach and Berends, 1965), might form a single hydrogen bond with guanine, but probably not with adenine. These two arrangements appear in Figure 11-1, and while neither is presented with any serious supposition that it is ever actually involved in the production of transversions each does illustrate a type of interaction which might be important.

12. Frameshift mutations

The fundamental hypothesis of Freese (1969*a*, *c*) that point mutations could be divided into transitions and transversions was prompted by the appearance of two distinct classes of mutants. These classes mapped at different genetic sites (Brenner, Benzer, and Barnett, 1958; Benzer, 1961*a*), and while one class was both induced and reverted by base analogues but not by proflavin, the other class was induced by proflavin, but was not induced to revert by base analogues (Freese, 1959*a* and *c*; Orgel and Brenner, 1961). (The structures of proflavin and of other mutagenic acridines appear in Figure 12-5.) It seemed most remarkable, however, that *none* of the mutants induced by base analogues occupied sites in common with mutants induced by proflavin, since both transitions and transversions should be able to occur at the same site. Two other observations made at about the same time also suggested the need for a refinement of Freese's taxonomy. First, studies of acridine binding to DNA (Lerman, 1961; Luzzati, Masson and Lerman, 1961) indicated that the acridine residues became sandwiched between adjacent base pairs, suggesting that during replication they might force the addition or deletion of a base pair. (This explanation of acridine mutagenesis now seems to be incorrect.) Second, it was observed by the bacteriophage workers at Cambridge University that mutations such as *o* and *h* which affect the proteins of the T4 particle itself were neither induced nor reverted by acridines. Since it was already clear that amino acid substitutions could frequently produce functional proteins, the putative transversions induced by acridines should have been frequently revertible by base analogues. Brenner, Barnett, Crick, and Orgel (1961) therefore proposed that mutants of the acridine type arise from the addition or deletion of a base pair.

The Cambridge workers had at this time become intrigued with a hypothetical relationship between translation and intragenic suppression. They supposed that messenger RNA might fold back upon itself to form a double-stranded structure, and that elements of any given codon are contained on both of the strands. As a result, they further imagined that a lesion at one end of a cistron might be suppressed by another lesion at the other end, the two lesions interacting directly in the messenger RNA. Centrally located lesions, on the other hand, would be suppressed by other centrally located lesions. To test this now defunct model, they examined revertants of T4rII mutants to see which ones reverted by means of intragenic suppressors, and where such suppressors would be located in relation to the original mutational site. While their results failed to support their original hypothesis, they did reveal that suppressors of mutations in the left end of the B cistron tended to be more distant from the suppressed sites, than were suppressors of mutations in the middle or the right end of the cistron. Later, when the remarkable properties of the mutant *r*1589 were observed (Champe and Benzer, 1962*a*), it became clear that the left end of the *r*IIB cistron is remarkably plastic, in the sense that it tolerates most amino acid substitutions and is, in fact, entirely dispensable. At the time, however, the left end of the B cistron was clearly the most convenient region in which to study mutual suppression by frameshift mutations.

A theoretical analysis of the possible relationships between the translation process and the reversion of acridine type mutations led the Cambridge group to the surprising conclusion that the analysis of frameshift mutations might reveal the fundamental nature of the genetic code. In particular, as we shall see, certain frameshift mutations which failed to show mutual suppression in pairwise combinations should succeed in higher order combinations, and this higher order should reveal the number of bases in the codon. Since a beginning to the experimental solution of the nature of the genetic code suddenly seemed to be in sight, a study of the reversion of frameshift mutations was initiated. Just at this time, however, Nirenberg reported at the Fifth International Congress of Biochemistry in Moscow that the *in vitro* synthesis of polyphenylalanine was directed by a polyuridylic acid template (see Nirenberg and Matthaei, 1961). Spurred on by the unexpected advance from the biochemists' corner, the Cambridge group worked at great speed to assemble a description of the fundamental properties of frameshift mutations. Their theoretical proposal (Brenner et al., 1961) was submitted in December of 1960; the report of their experimental results and the ensuing analysis (Crick, Barnett, Brenner, and Watts-Tobin, 1961) made its appearance on the last working day of 1961.

GENETIC PROPERTIES OF
FRAMESHIFT MUTATIONS

The elementary algebra of frameshift mutations, deduced from patterns of mutual suppression, has already been described in Chapter 8; we will concentrate here on additional properties of the mutants. In order to facilitate discussion, a map of the T4rIIB mutants studied by the Cambridge workers is shown in Figure 12-1 (adapted from Barnett et al., 1967). Multiple occurrences of many of these mutants were recorded, but have been omitted from the figure in order to achieve simplicity. This map was first constructed by the process of "mutational leapfrog." A T4rII mutant (*r*P13) originally induced by proflavin was renamed *r*FCO, and revertants were obtained from it. These were backcrossed to the wild type in order to isolate different suppressors of *r*FCO, each suppressor being an

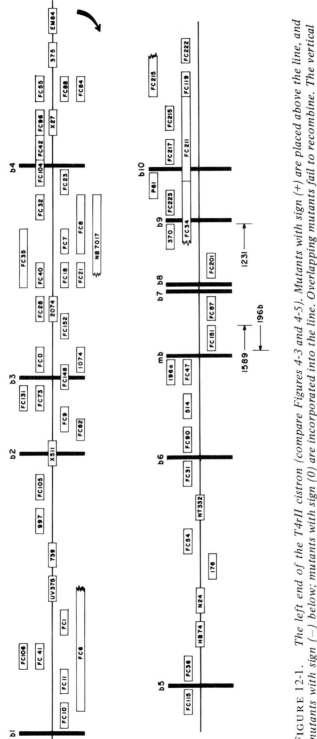

FIGURE 12-1. *The left end of the T4rII cistron (compare Figures 4-3 and 4-5). Mutants with sign (+) are placed above the line, and mutants with sign (−) below; mutants with sign (0) are incorporated into the line. Overlapping mutants fail to recombine. The vertical bars represent the barriers. The endpoints of several large mapping deletions are indicated at the bottom. The mutants are arranged in the correct order, but the map only approximately summarizes quantitative recombination frequencies. Data from Barnett et al. (1967).*

rII mutant itself. A few of these suppressors were then used as sources of revertants from which second-order suppressors were obtained by means of backcrosses to wild type. By iteration, most of the mutants of Figure 12-1 were eventually collected. FCO was arbitrarily assigned sign (+), which would correspond to a reading frame shift to the left (see Figure 8-3), and all other mutants were assigned signs according to their descent from FCO.

Operationally, frameshift mutants are conveniently defined in terms of their susceptibilities to induction and reversion by specific mutagens. In bacteriophage T4, frameshift mutations are induced by acridines, and are nearly always induced to revert by acridines (but never by base analogues), whereas base pair substitution mutations are induced to revert by base analogues, and are not induced to revert by acridines. Similar rules apply in bacteria, except that chemically more complicated types of acridines must be used (see below). In all cases, care must be taken to avoid the complications of extracistronic suppression (see Chapter 17). A small number of exceptions to the rule that no mutants are reverted by both acridines and by base analogues have been described for the T4rII system (Orgel and Brenner, 1961; Krieg, 1963a and b). However, the proportion of such mutants is very small, and while they usually revert to an apparently fully functional state when treated with base anologues, they revert to a partially defective state when treated with acridines. It is not even clear whether or not these revertants may contain extracistronic suppressors. In addition, it is to be expected that additions or deletions of exactly three bases should very occasionally erase base pair substitution lesions, and *vice versa*. The exceptional rII mutants therefore pose no serious threat to the theory of frameshift mutations.

It is obvious from the algebra of frameshift mutations (Chapter 8) that if N mutants of like sign are combined, mutual suppression may occur when N equals the number of bases in a codon. In the T4rII system, the wild phenotype is recovered when $N = 3$, for instance with the $(- - -)$ combination rFC10, rFC21, rFC23 and with the $(+ + +)$ combination rFC106, rFC0, rFC40 (Barnett et al., 1967). Combinations of four and five mutations always exhibited the mutant phenotype. However, a combination of six lesions again exhibited the wild phenotype (rFCO, rFC40, rFC55, rFC36, rFC31, rFC47). These results demonstrated for the first time that codons contain three bases; the logical possibility that codons actually contain a multiple of three bases was very unlikely, as was the possibility that most codons contain three bases, but that a few contain some other number.

Initially, it was frequently supposed that frameshift mutations consisted of additions or deletions of single base pairs, although the 1961 report of Crick et al. indicated the possibility of more complicated lesions. The fine-scale mapping of the left end of the rIIB cistron revealed that many frameshift mutants failed to recombine with two or three or more mutants which were themselves capable of mutual recombination (Figure 12-1). The dimensional complexity of these mutants was revealed by a study of their abilities to become incorporated into heteroduplex heterozygotes (Drake, 1966*a*). Crosses were performed between various frameshift mutants and a nearby base pair substitution (ochre) mutant; at the same time, DNA replication was inhibited by FUDR in order to accumulate heteroduplex heterozygotes (see Chapter 3). The *r* mutations present in heterozygotes were then determined; generally only a single *r* mutation was observed in a single heterozygote. Frameshift mutations with sufficiently small lesions should be recovered from such heterozygotes just as frequently as is the ochre mutation, whereas "large" frameshift lesions would be recovered relatively rarely. A smooth distribution of recovery frequencies was observed (Figure 12-2), and the size of a frameshift mutation estimated from its characteristic heterozygote frequency was well correlated with its size estimated from the number of genetic sites which it covered. Some of the largest frameshift mutations were estimated to contain deletions of about 20 base pairs. A few mutants (*r*FC215, *r*FC7, *r*FC36, *r*FC223) occupied only a single gentic site, but appeared "large" in the heterozygote test; such mutants probably contain extensive base pair additions.

   The combination of a pair of frameshift mutations of unlike sign produces a normal reading frame outside of the interval between the two mutants; however,

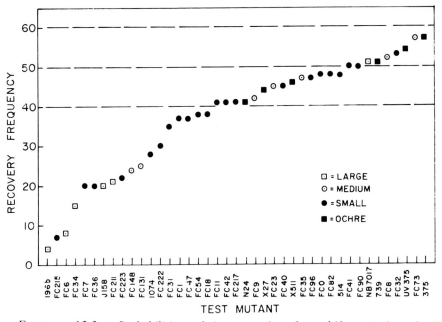

FIGURE 12-2. *Probabilities of incorporating frameshift mutations into recombinational heterozygotes. "Large" and "small" symbols refer to the extent of the lesion on the genetic map (Figure 12-1), while mutants towards the left of the figure are "large" in the physical sense because many base pairs have been added or deleted; r196b is a mapping deletion. Most of these heterozygotes are of the heteroduplex type, except at the extreme left, where terminal redundancy heterozygotes may predominate.*

the interval between them contains a shifted reading frame, and therefore contains a series of amino acid substitutions (see Chapter 8 and below). As a result, two types of restrictions are placed upon the functioning of (+ −) or (− +) combinations. First, none of the codons which appear in the shifted frame may be chain terminating. Second, none of the amino acid substitutions which occur may be deleterious in themselves; this condition appears to be usually satisfied in the special case of the left end of the rIIB cistron, but probably not in many other parts of the rII region, where frameshift mutations often revert by means of extremely closely linked suppressors or by repair of the original lesion (Drake, unpublished results). For example, the insertion of a proline residue into the middle of an α-helical segment, or the introduction of a cysteine where it might form a disulphide bridge, might well be disasterous to an otherwise relatively plastic region of the protein.

Specific combinations of (+) and (−) mutants in the rII system often fail to achieve mutual suppression. Such incompetent combinations arise when the two mutants span a few specific points along the map, these points being unidirectional: a (+ −) combination across one of them may acheive suppression where a (− +) combination fails. The points which appear to prevent mutual suppression are called "barriers" (Crick et al., 1961).

A number of barriers have been studied in detail (Barnett et al., 1967). Of the barriers preventing reading frame shifts to the left, several are clearly due to chain-terminating codons generated within the shifted reading frame. These barriers disappear when the (+ −) combination is plated on cells permissive for chain-terminating mutants, and are also induced to "revert" by base analogue mutagenisis. Specifically, barriers **b2**, **b5**, and **b6** appear to contain UGA codons; barrier **b9**, an amber codon; and barriers **b3**,, **b4**, and either **b7** or **b8**, ochre codons. Barriers **b1** and **b10** remain uncharacterized, but probably do not contain chain-terminating codons. Barrier **b7** or **b8** (whichever does not contain the ochre codon) is a leaky barrier, and is therefore probably the result of an unacceptable amino acid. The barriers preventing reading frame shifts to the right have not been studied in detail, but one, the "minute" barrier, **mb**, may also result from an unacceptable amino acid, since (− +) combinations across it produce tiny plaques on K(λ) cells. In view of the generally permissive nature of the left end of the rIIB cistron, the nature of the barriers which do not appear to contain chain-terminating codons remains mysterious.

A reading frame shift, in addition to creating chain-terminating codons, can also erase them (Brenner and Stretton, 1965). The following combinations, for instance, all grow on nonpermissive host cells: rFC6, rEEM84, rFC84 (− am +); rFC0, r2074, rFC23 (+ am −); and rFC10, rUV375, rFC0 (− och +).

An array of adjacent cistrons controlled by a single regulatory site at one end, and cotranscribed onto a single polycistronic messenger RNA molecule, is called an *operon*. Mutational lesions which introduce chain-terminating codons into one of the cistrons often result in the partial inactivation of all cistrons on the side of the lesion away from the regulatory (operator) site. Such mutants are said to be *polar* (see Chapter 18). Polarity is also commonly observed to result from frameshift mutations, for instance, in the *E. coli lac* operon (Malamy, 1966) and in the *S. typhimurium his* and *tryp* operons (Martin et al., 1966; Bauerle and Margolin, 1966*b*). The polarity exhibited by frameshift mutations probably results from chain-terminating codons which appear in the shifted reading frame.

An anomalous pattern of mutual suppression was observed with certain (+ +) combinations in the T4*r*II system (Barnett et al., 1967). A few such combinations resulted in partial restoration of the *r*IIB function. The restoration was poor, however, the double mutants producing minute plaques on K(λ) cells and the *r* phenotype on B cells. Furthermore, the effect was restricted to pairs of (+) mutants spanning a few points on the map, particularly barrier **b6** and barrier **b7/b8**. The correlation between barriers and sites permitting (+ +) combinations to score as minutes was greatly strengthened by the observation that after reversion of barrier **b6** by base analogue mutagenesis, (+ +) double mutants spanning this barrier no longer grew at all on K(λ) cells. However, only particular chain-terminating barriers tend to favor the appearance of (+ +) minutes. Barnett et al. argued that (+ +) minutes probably arise from occasional errors of translation which effectively introduce an additional frameshift consisting of sign (+). It is assumed that reading interrupted by a chain-terminating codon might occasionally start up again in a new reading frame, either by continued growth of the nascent polypeptide chain, or by reinitiation of an entirely new chain. Sarabhai and Brenner (1967) have presented evidence strongly suggesting that out-of-phase reinitiation may occur near an *r*II chain-terminating codon.

It is often convenient to suppose that base pair substitutions produce mutants of sign (0), for instance, when considering the erasure of a chain-terminating codon within a frameshifted region. Similarly, deletions which behave as missense mutations in the *r*1589 test also possess sign (0). Since nearly all mutants of sign (0) in the left end of the *r*IIB cistron are chain terminators, this region is apparently indifferent to missense mutations. This being the case, it would be expected that the chain-terminating mutants in this region could be reverted by additions or deletions of three base pairs. This possibility has not been very seriously tested, however.

EFFECTS ON TRANSLATION

Base pair substitutions are expected to cause either single amino acid substitutions or chain termination. A pair of mutually suppressing frameshift mutations, however, are expected to cause a more extensive alteration of the amino acid sequence (Crick et al., 1961). Depending upon the geometry of the sites themselves, two effects should be observable: first, within the frameshifted interval, the entire amino acid sequence may be altered; and second, the net number of amino acids may also be altered. Shortly after the genetic properties of the frameshift mutants were determined, Streisinger and his associates initiated a study of the amino acid sequence of T4 lysozyme. By 1966, they were able to demonstrate the occurrence of drastically altered amino acid sequences resulting from frameshift mutations. A number of combinations of mutually suppressing mutations have now been analysed by this method (Streisinger et al., 1966; Terzaghi et al., 1966; Okada et al., 1966; Inouye et al., 1967; Lorena, Inouye, and Tsugita, 1968). Similar examples have also been reported within the *E. coli* tryptophan synthetase A protein (Brammar, Berger, and Yanofsky, 1967).

The analytical procedure consists of determining the amino acid sequence in the affected region for both the wild type and the double mutant, and then deducing the associated mRNA codons. Despite the extensive codon degeneracy, it has been possible to find unique codon solutions for most of the amino acids, primarily because certain members of a given group of bases are often involved in coding for two different amino acids. The mRNA base sequence can be deduced by writing down all possible codons for each amino acid, and looking for a frameshift which reveals the relationship between the wild type and mutant codons. The enthusiastic reader may wish to tackle one of these problems before examining Figure 12-3, where several sets of data are analysed. The double mutant *e*J42*e*J44 induces the following conversion: wild type = . . Thr . . Lys . . Ser . . Pro . . Ser . . Leu . . Asn . . Ala . . ; double mutant = . . Thr . . Lys . . Val . . His . . His . . Leu . . Met . . Ala . .

Four examples drawn from the lysozyme system are illustrated in Figure 12-3. In terms of base additions and deletions, these examples comprise the following combinations: $(-1 + 1)$, $(-2 + 2)$, $(+2 + 1)$, and $(-1 - 1 + 2)$. They are therefore examples of single base additions and deletions, of double base additions and deletions, of a net change in the number of amino acids, and of mutual suppression by three mutations of identical sign, respectively.

In addition to confirming the geneticists' predictions, these results hint at an unanticipated phenomenon. In many cases, the observed base change either subtracted from a region of local base redundancy, or else created such redundancy. The *e*J17 mutation, for instance, inserted UG or GU at one end of the three internucleotide positions associated with . . GU . . ; the *e*J42 mutation deleted one base from the . . XAApuA . . sequence; and the *e*JD5 mutation deleted XG or GA from the . . XGAA . . sequence. This effect will be considered in more detail below.

Frameshift mutations clearly add and delete variable numbers of bases; when the number is three or a multiple thereof, the mutation will have sign (0), and although it may or may not inactivate the corresponding protein, it will not induce a reading frame shift. A few mutationally altered proteins containing amino acid deletions rather than substitutions have been observed, including human hemoglobins with deletions of both single amino acids (Jones et al., 1966; de Jong, Went, and Bernini, 1968) and multiple (five) amino acids (Bradley, Wohl, and Rieder, 1967); the second example is particularly interesting because *leu, his, cys, asp,* and *lys* were removed from a region whose sequence

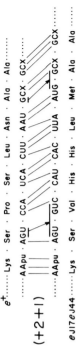

(−1+1)

e⁺ ...... Thr · Lys · Ser · Pro · Ser · Leu · Asn · Ala ......
...... ACX · AApu · AGU · CCA · UCA · CUU · AAU · GCX ......
...... ACX · AApu · AGU · GUC · CAU · CAC · UUA · UUA · AUG · GCX ......
eJ42 eJ44 ...... Thr · Lys · Val · His · His · Leu · Met · Ala ......

(+2+1)

e⁺ ...... Lys · Ser · Pro · Ser · Leu · Asn · Ala · Ala ......
...... AApu · AGU · CCA · UCA · CUU · AAU · GCX · GCX ......
...... AApu · AGU · GUC · CAU · CAC · UUA · AUG · GCX · GCX ......
eJ17 eJ44 ...... Lys · Ser · Val · His · His · Leu · Met · Ala · Ala ......

(−1−1+2)

e⁺ ...... His · Leu · Leu · Thr · Lys · Ser · Pro · Ser ......
...... CApy · pyUX · pyUX · ACpu · AApu · AGU · CCA · UCA ......
...... CApy · pyUX · CAA · AAG · UGU · CCA · UCA ......
eJD10 eJ42 eJ17 ...... His · Leu · Gln · Lys · Cys · Pro · Ser ......

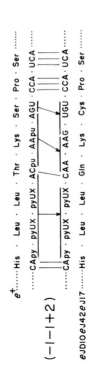

(−2+2)

e⁺ ...... Asp · Thr · Glu · Gly · Tyr · Tyr · Thr · Ilu ......
...... GApy · ACX · GAA · GGpy · UApy · UApy · ACX · AUX ......
...... GApy · ACX · AGG · pyUA · pyUX · CApy · ACX · AUX ......
eJD5 eJ201 ...... Asp · Thr · Arg · Leu · Leu · His · Thr · Ilu ......

FIGURE 12-3.   *Amino acid replacements associated with mutually suppressing frameshift mutations in the T4 lysozyme cistron. For each example, the wild type amino acid sequence is given at the top, the multiple mutant sequence at the bottom. The corresponding messenger RNA sequences appear in the middle. The numbers in parentheses to the left of each example indicate the numbers of bases added or deleted by each mutation in the series; the locations of the mutations are indicated by the arrows, as far as they can be pinpointed. X stands for "any base" (or for U, C, or A at the Ilu positions in the special case of the (−2 +2) combination); py stands for "any pyrimidine" and pu for "any purine."*

suggests local redundancy: ... *leu . his . cys . asp . lys . leu . his* .... These mutations were interpreted as ordinary small deletions which arose from unequal crossing over (see Chapter 9). If frameshift mutations and true macrolesions arise by different mechanisms, however, then there is a strong possibility that many small deletions represent frameshift mutations of sign (0).

THE INDUCTION OF
FRAMESHIFT MUTATIONS

A great variety of chemical and physical treatments induce base pair substitutions in various organisms, but rather few treatments induce significant numbers of frameshift mutations. About half of the T4*r*II mutants induced by ultraviolet irradiation contain frameshift lesions (Drake, 1963*b* 1966*c*). Certain steroidal diamines induce revertants of *r*II frameshift mutants (Mahler and Baylor, 1967). The mutagenicity of proflavin was first observed in bacteriophage T2 by DeMars (1953), and today a wide range of acridines (Figure 12-5) is known to induce frameshift mutations in bacteriophages T2 and T4 (Orgel and Brenner, 1961; Lerman, 1964*b*). Like the base analogues, the acridines are only mutagenic to bacteriophages during their intracellular phase, at least in the absence of white light irradiation. However, proflavin, which is among the most powerful of the mutagenic acridines, is not an effective mutagen for bacteriophage λ and may in fact only act under somewhat special conditions. For instance, infection by the T-even bacteriophages induces a greatly increased passive permeability of proflavin in the host cells (Silver, 1965, 1967). In addition, both the high incidence of genetic recombination, and the random distribution of chromosomal ends around the map, may promote the mutagenicity of proflavin in the T-even bacteriophages (see below).

Proflavin is usually not an effective mutagen for bacteria (Lerman, 1963; Orgel, 1965), but it may become very mutagenic if applied to *E. coli* during sexual conjugation (Sesnowitz-Horn and Adelberg, 1968). In a few instances, acridines have been reported to induce mutations under normal growth conditions (Witkin, 1947; Eisenstark and Rosner, 1964; Zamperi and Greenberg, 1965; Sicard, 1965; Stewart, 1968). Since acridines can also induce mutations photodynamically (Webb and Kubitschek, 1963; Ritchie, 1964, 1965), and since such mutations contain transitions and transversions, but rarely or never frameshift lesions (Drake and McGuire, 1967*b*), care must be taken to avoid white light side effects when testing the mutagenicity of acridines. The frequency of mutations induced in the *E. coli* system, however, was not increased by deliberate exposure to visible light (Greenberg, personal communication). On the other hand, in none of these cases have the induced mutants actually been shown to contain frameshift lesions, and many of the *lac* mutants are in fact revertible by 2-aminopurine (Greenberg, personal communication).

Mutations are induced in enteric bacteria by a group of agents (the ICR compounds) which consist of an acridine moiety to which is attached a side chain, often an alkylating agent (see Figure 12-7 and Table 12-2). These compounds were synthesized by Creech and his associates at the Institute for Cancer Research, and many of the mutations which they induce clearly contain frameshift lesions. For instance, Ames and Whitfield observed that the mutations induced by ICR-191 were reverted by the same agent, but were not reverted by base analogues or by nitrosoguanidine, a powerful mutagen which appears to induce many types of base pair substitutions. All of the mutants were polar, and none of them reverted by means of "external" (extracistronic) suppressors; polar chain-terminating mutants, on the other hand, frequently revert by means of extracistronic suppressors. In addition, Sederoff (quoted in Ames and Whitfield, 1966) observed that ICR-170 induced revertants of T4rII frameshift mutants, and Strigini (personal communication) observed that ICR-191 induced revertants of T4r frameshift mutants, but not of a transition mutant. Finally, mutants specifically identified in the *E. coli* tryptophan synthetase A protein as frameshift lesions were induced to revert by ICR-191A (Brammer, Berger, and Yanofsky, 1967), and *S. typhimurium his* mutants which were induced to revert by ICR-191 acquired nearby suppressors, the original and suppressor lesions having many of the properties to be expected of (+) and (−) mutants (Martin, 1967).

It should be noted that compounds of the ICR series sometimes appear to induce revertants of base pair substitution mutants (Ames and Whitfield, 1966; Whitfield, Martin, and Ames, 1966; Brammar, Berger, and Yanofsky, 1967; Berger, Brammar, and Yanofsky, 1968). Some but not all of these revertants may well contain extracistronic suppressors, since such suppressors sometimes do arise by means of frameshift mutations (Magni, von Borstel, and Steinberg, 1966; Fink, Klopotowski, and Ames, 1967), probably in sRNA genes (see Chapter 17). In addition, some mutants induced by ICR compounds may be revertible by alkylating agents (Malling, 1967; Oeschger and Stahl, 1967). The most reliable criteria for identifying bacterial frameshift mutations by means of reversion tests therefore, consist of a positive response to an ICR compound and a negative response to DES and/or nitrosoguanidine.

A number of instances of putative frameshift mutagenisis in eukaryotic organisms have been claimed; the supporting evidence is roughly inversely proportional to the phylogenetic standing of the organism studied. In the yeast *Saccharomyces cerevisiae,* two different types of mutational lesions can be observed (Magni, 1963; Magni, von Borstel, and Sora, 1964; Magni and Puglisi, 1966). One type appears preferentially during mitosis, and frequently reverts by extracistronic suppression; these mutations probably result from base pair substitutions. The other type appears preferentially during meiosis and is correlated with genetic recombination; forward mutations of this type do not revert by external suppressors, but are induced to revert by 5-aminoacridine and by ICR-170. These mutations clearly contain frameshift lesions.

Mutations induced in the fungus *Neurospora crassa* by ICR-170 have also been interpreted as frameshift lesions (Brockman and Goben, 1965; Malling, 1967). These mutants are rarely leaky, and rarely exhibit intracistronic complementation (see Chapter 18), nor do they contain large deletions. A number of them are induced to revert by ICR-170, but not by nitrous acid or by the alkylating agent EMS. Acridine has also been reported to produce mutations in the fungus *Aspergillas nidulans* (Ball and Roper, 1966), but in this case all that can be said is that the process appears to be distinct from photodynamic mutagenesis. Finally, mutations induced in the fruit fly *Drosophila* by ICR-170 have been surmised to contain frameshift lesions because of their different properties from mutants induced by X-rays and alkylating agents, and in particular because they were never temperature-sensitive (Carlson, Sederoff, and Cogan, 1967). Considerable additional information, however, is obviously needed before frameshift mutations can be unequivocally identified in these systems.

THE INTERACTION OF ACRIDINES WITH DNA

There are two types of interactions between acridines and double stranded DNA, characterized by strong and by weak binding, respectively (Peacocke and Skerrett, 1956). Strong binding is observed at acridine/base pair ratios below about 0.2 or 0.4; weak binding, which occurs essentially on the "outside" of the DNA molecule, is observed at dye/base pair ratios from 0.4 to 1. Lerman (1961) proposed that strong binding occurs by intercalation of acridines between adjacent base pairs (Figure 12-4). The evidence to support this model is very strong. The flow dichroism and flow-polarized fluorescence of bound acridine residues indicate that the plane of the acridine ring is perpendicular to the long axis of the DNA molecule Lerman, 1963). The amino groups of proflavin are protected against diazotization, and the hydrogen-bonded amino groups of the DNA bases retain their inability to react with formaldehyde (Lerman, 1964*a*). Each intercalated acridine residue should increase the length of the DNA molecule by the equivalent of one base pair (3.4 Å), and at the same time increase its rigidity, both by stacking interactions and by stretching of the backbone. Dye binding does increase the viscosity and decrease the sedimentation coefficient of DNA (Lerman, 1961; Drummond et al., 1966); it also decreases the mass per unit length of DNA, and increases its radius of gyration (Luzzati, Masson, and Lerman, 1961; Mauss et al., 1967). The elongation of bacteriophage T2 DNA by proflavin has been autoradiographically visualized by Cairns (1962).

If intercalation occurs primarily between base pairs, then thermal denaturation of the DNA might largely eliminate strong binding. However, since it does not (Drummond, Simpson-Gildemeister, and Peacocke, 1965), Pritchard, Blake, and Peacocke (1966) have suggested that acridines are intercalated between single adjacent (noncomplementary) bases. On the other hand, even denatured DNA contains appreciable amounts of double-stranded structure. Their model also purports to explain the ionic effect observed upon strong dye binding, since

FIGURE 12-4.   *Schematic model for the intercalation of acridines (in black) into DNA; after Lerman (1946b).*

the acridine ring nitrogen could then interact with the DNA phosphate group. However, acridines intercalated in this way might well decrease rather than increase the viscosity of DNA, since they might introduce kinks into the molecule.

The strong binding of acridines to DNA was at one time believed to exhibit a preference for intercalation next to A:T base pairs (Kleinwächter and Koudelka, 1964; Tubbs, Ditmars, and Van Winkle, 1964). A recent reinvestigation of this effect, however, revealed that no such preference exists (Chan and Van Winkle, 1969).

Quite a few acridines have been tested for mutagenicity (Orgel and Brenner, 1961; Lerman, 1964b); these are listed in Table 12-1, where Richter's ring numbering system is used (Figure 12-5). (The reader should be warned that while much but not all of the British literature uses this system, much but not all of the American literature uses Graebe's system, also shown in Figure 12-5; sometimes one can only guess which system is used.) Attempts to discover consistent correlations between mutagenicity and ring substitutions are discouraging (Riva, 1966). The more basic compounds in the aminoacridine and diaminoacridine series roughly tend to be more mutagenic in the sense that the optimally mutagenic concentration is lower, but the optimal mutagenesis itself does not correlate particularly well with basicity. Methylation of the 10-nitrogen may abolish mutagenicity altogether, except in the case of acriflavin (Lerman, 1964b). Even the exact configuration of the acridine ring itself is not required; the very similar phenanthridines (Figure 12-5) are also mutagenic. The

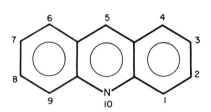

ACRIDINE: RICHTER SYSTEM

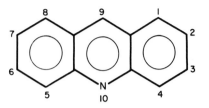

ACRIDINE: GRAEBE SYSTEM

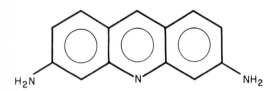

PROFLAVIN

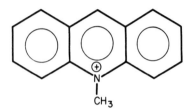

IO-METHYLACRIDINIUM

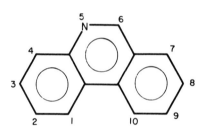

PHENANTHRIDINE

FIGURE 12-5.   *Acridine and phenanthridine ring numbering systems; the Richter system is used in the text.*

interpretation of comparitive mutagenicities is hideously complicated by uncertainties about the extent to which the compounds penetrate the cell membrane and interact with the DNA (although Lerman [1964*b*] showed that the 10-methylated acridines could penetrate the cell), and by the unknown role of various experimental conditions. Drake observed that the proflavin-induced reversion of T4*r*UV58 was about 10-fold more effective in broth at or above pH 7.6 than at or below pH 7.2. Lerman (1964*b*) found acridine very much less mutagenic than proflavin when measuring the induced reversion of T4*r*P3, whereas Orgel and Brenner (1961) generally found the two compounds to be very similar when tested with a variety of tester mutants (in both cases using near-optimal mutagen concentrations).

The mutagenically significant interaction of acridines with T4 DNA may be a complex process, since the mutational response shows a dye concentration

TABLE 12-1

Compound	Basic pK	Optimal Mutagenic Concentration (μg/ml)	Mutagenicity
Acridine	5.3	90	Moderate
1-Aminoacridine	4.2	150	Weak
2-Aminoacridine	7.7	16	Strong
3-Aminoacridine	5.6	80	Weak
4-Aminoacridine	5.7	80	Strong
5-Aminoacridine	9.6	8	Strong
2,5-Diaminoacridine	11.1	8	Strong
2,7-Diaminoacridine	7.8	16	Moderate
2,8-Diaminoacridine (proflavin)	9.3	5	Strong
10-Methylacridinium			None
5-Amino-10-methylacridinium			None
5-(N-piperidino)-acridine			Moderate
2,8-diamino-10-methylacridinium (acriflavin)			Strong
2,8-diamino-3,7-dimethylacridine (acridine yellow)	9.8	2	Strong
2,8-bisdimethylaminoacridine (acridine orange)	10.1	20	Weak
Phenanthridine			Moderate
1-Aminophenanthridine			Weak
6-Aminophenanthridine			Moderate

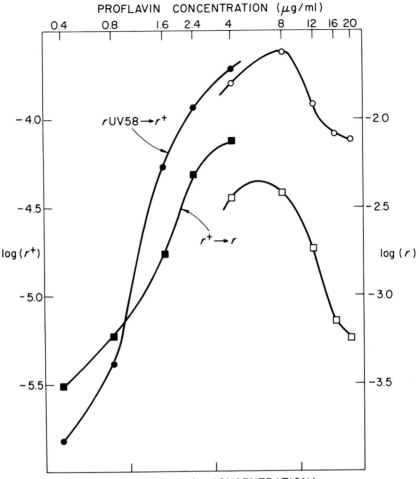

FIGURE 12-6.   *The concentration dependence of proflavin mutagenesis in bacteriophage T4. E. coli BB cells at 37° in M9 medium were infected with an average of 5 particles at 0 and at 4 minutes. Proflavin was added to the indicated concentrations at 8 minutes, and the complexes were diluted out of the proflavin at 40 minutes. Lysis was completed with chloroform at 90 minutes, and mutant frequencies were measured.*

power dependence greater then unity. Figure 12-6 shows a plot of log M against log C, where M is the frequency of induced mutants observed with proflavin at concentration C. Forward mutations appear at approximately the 1.4 to 2.6 power of proflavin concentration (estimated from the responses at the 4.0/0.4 and 2.4/1.6 dye concentrations, respectively), while revertants of rUV58 appear at approximately the 2.1 to 3.4 power of proflavin concentration (estimated from the responses at the 4.0/0.4 and 1.6/0.8 dye concentrations, respectively).

Lerman (1964*b*) also mentions observing a 1.9 order reversion response using 5-aminoacridine. Both Lerman and Drake observed decreases in the mutagenic responses at supraoptimal dye concentrations. All of these data must be interpreted cautiously, however, because of the obviously complex nature of the responses as illustrated in Figure 12-6, and because of uncertainties concerning the intracellular dye concentrations.

Ames and Whitfield (1966) have examined the comparitive mutagenicities of several compounds of the ICR series (Figure 12-7 and Table 12-2) by measuring the induced rate of reversion of a frameshift mutation in the *S. typhimurium his* C gene. Compounds whose side chains end in chloro groups proved to be stronger mutagens than compounds whose side chains end in hydroxyl or amino groups, suggesting that alkylation might strongly promote mutagenicity. However, alkylating agents themselves are not generally able to induce frameshift mutations; therefore, the mutagenic action of these compounds depends upon the rings. Since the nonalkylating compounds are usually at least weakly mutagenic, alkylation appears to be a secondary aspect of frameshift mutagenesis by ICR compounds. It is interesting to note that the widely used antimalarial compound quinacrine is detectably mutagenic in these tests. It will obviously be desirable to investigate the nature of the interactions of these compounds with DNA.

THE MECHANISM OF FRAMESHIFT MUTATION

Three models have been offered to explain the origin of frameshift mutations: errors of replication, errors of recombination due to unequal crossing over, and errors of DNA repair. Each model assumes that the mutagenicity of acridines results from intercalation.

Brenner et al. (1961) suggested that intercalation between adjacent bases (as opposed to intercalation between adjacent base pairs) could induce miscopying errors which would add or delete single bases. Since inhibition of T4 DNA synthesis by FUDR (sufficient to obliviate base analogue mutagenesis) reduced proflavin mutagenesis only slightly (Drake, 1964), and since many frameshift mutations clearly add or delete two or more base pairs (Drake, 1966*a*; Streisinger et al., 1966), frameshift mutations are unlikely to arise as simple errors of replication.

Lerman (1963) suggested that acridines induce unequal recombination (unequal crossing over) by forcing mistakes in homologous pairing. However, this model does not clearly explain the variable number of base pairs added or deleted in frameshift mutations, and is difficult to reconcile in detail with models of recombination which suppose that homologous pairing first involves single-stranded, and then hybrid intermediates (Figure 3-4). The model does suggest that correlations should exist between recombination and frameshift mutation, which would explain the strong response of T-even bacteriophages to acridines, compared with the response of bacteriophage λ and bacteria, neither of which experience recombination very frequently. [The exceptional muta-

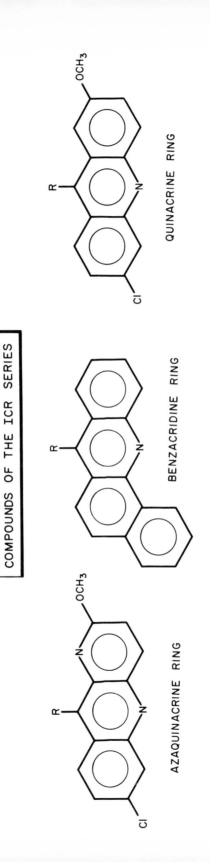

FIGURE 12-7. *Ring moieties of the ICR compounds.*

TABLE 12-2

MUTAGENICITIES OF THE ICR COMPOUNDS*

Side Chain (R)	Ring		
	Azaquinacrine	Benzacridine	Quinacrine
$-NH-CH_2-CH_2-CH_2-NH-CH_2-CH_2-Cl$	ICR-372 (strong)	ICR-370 (strong)	ICR-191 (strong)
$-NH-CH_2-CH_2-NH-CH_2-CH_2-Cl$	ICR-364 (weak)	ICR-312 (weak)	ICR-171 (weak)
$-NH-CH_2-CH_2-CH_2-N\begin{smallmatrix}CH_2-CH_3\\ \\CH_2-CH_2-Cl\end{smallmatrix}$	ICR-340 (strong)	ICR-292 (weak)	ICR-170 (weak)
$-NH-CH_2-CH_2-CH_2-NH-CH_2-CH_2-OH$	ICR-372-OH (weak)	ICR-370-OH (positive)	ICR-191-OH (0)
$-NH-CH_2-CH_2-NH-CH_2-CH_2-OH$	ICR-364-OH (weak)		
$-NH-CH_2-CH_2-CH_2-NH-CH_2-CH_2-NH_2$	ICR-364-NH₂ (positive)		
$-NH-CH\begin{smallmatrix}CH_3\\|\end{smallmatrix}-CH_2-CH_2-CH_2-N\begin{smallmatrix}CH_2-CH_3\\ \\CH_2-CH_3\end{smallmatrix}$			Quinacrine (positive)

*Source: Ames and Whitfield (1966). The reversion of only a single frameshift mutation was studied, so mutational categories have been indicated only approximately. "Positive" responses are very weak.

genicity of acridines in T4 is not due to the glucosylation of the 5HMC residues (Drake, 1964).]

Impressive correlations have been established between frameshift mutation and recombination. In *Saccharomyces,* frameshift mutations arise spontaneously and are induced by acridines during meiosis (when recombination occurs), but rarely during mitosis (Magni, 1963; Magni, von Borstel, and Sora, 1964). Furthermore, the mutants are highly recombinant for outside markers, regardless of when they arise. In *E. coli,* proflavin becomes much more mutagenic during conjugation, and the resulting mutants arise in the vicinity of recombination events (Sesnowitz-Horn and Adelberg, 1968). In *B. subtilis*, however, no such correlation has been observed (Stewart, 1968). In bacteriophage T4, Strigini (1965) observed an apparent correlation between frameshift mutation and recombination, but these results exhibited certain unexpected complications which will be considered below.

A number of tests of Lerman's unequal recombination model have been performed with the T4*r*II system. For example, since ultraviolet irradiation sharply increases recombinant frequencies in T4 crosses, its effect upon proflavin mutagenesis was tested; yet no change was observed (Drake, 1964). Also, since reciprocal crossing over could simultaneously produce two different mutations, the composition of mutant clones was examined; all were homogeneous (Drake, 1964). Neither of these tests was rigorous, however, especially since recombination generally appears to be nonreciprocal in bacteriophages (see Chapter 3). In another test, Brenner (personal communication) and Lindstrom and Drake (unpublished data) have compared reversion of a frameshift mutation with recombination of a pair of base pair substitution markers, one on either side of the frameshift mutation. The outside markers were conditional lethals, and could be scored independently of revertants of the frameshift mutation. Reversion was not correlated with recombination. Drake, in still another experiment, examined the joint reversion of pairs of frameshift mutants in the T4*r*II region. Wild type revertants appeared at the frequency expected of two independent events, whereas if reversion were triggered by nearby recombination events, simultaneous reversion of both lesions might frequently have occurred. In other unpublished experiments, Drake examined the reversion of a frameshift mutation when recombination was depressed in its immediate vicinity because most of the chromosomes with which it would recombine carried a deletion of the same region (Drake, 1967); proflavin-induced reversion was not suppressed. It therefore appears that the correlation between recombination and frameshift mutation observed in certain organisms is largely absent from T4.

Strigini (1965) detected what appeared to be a direct correlation in T4 between frameshift mutation and recombination. In an experiment similar to that described above (except that the outside markers were completely outside of the *r*II region), he observed that newly arisen frameshift mutations were heterozygous for one or both of the outside markers. Drake (1964) had also observed that newly arisen *r*II mutants were r/r^+ heterozygotes. However, those observed by Strigini were extraordinarily long, often extending over 10 or more map units, whereas normal recombinational heterozygotes are only about 2 to 4 map units long (Berger, 1965). Unless the mutational heterozygotes are "pathological" recombinational heterozygotes, therefore, they are much more likely to be terminal redundancy heterozygotes, which are 20 to 40 map units long (Doermann and Boehner, 1963; Wiemann, 1965; MacHattie, Ritchie, Richardson and Thomas, 1967). If frameshift mutations arise in regions of terminal redundancy, but independently of recombination, they are nevertheless apt to be secondarily correlated with recombination, since nearby markers included in the terminally redundant region will be heterozygous about half the time simply because opposite ends of the chromosome are quickly randomized with respect to their alleles. In addition, the physical ends of T4 DNA molecules are themselves highly recombinogenic (Mosig, 1963; Doermann, and Parma, 1967).

Another effect of 5-aminoacridine upon T4 replication has recently been observed by Lerman and Altman (personal communication): the number of free ends among intracellular DNA molecules is greatly increased. Acridine mutagenesis might therefore occur simply because frameshift mutations normally arise at such free ends.

It was noted previously that frameshift mutations often appear to introduce base pair additions and deletions in the immediate vicinity of local base pair redundancies. As a result, Streisinger surmised that frameshift mutations may be the consequence of mispairing errors during the repair of single-strand interruptions in a double-stranded DNA molecule (Figure 12-8). He proposed that local melting away from such a break, followed by erroneous reannealing promoted by local base pair redundancy, would generate a frameshift mutation if the misannealed configuration were sealed up before the more probable, correctly annealed configuration was reestablished. Model building experiments suggest that the resulting single-strand loops could form without extensive local disorganization of the double helical structure of DNA (Fresco and Alberts, 1960). As illustrated in Figure 12-8, Streisinger's model generates both addition and deletion mutations. Inouye et al. (1967) and Tsugita et al. (1969) have presented specific models to explain the origins of several specific frameshift mutations.

The Streisinger model successfully explains a number of aspects of frameshift mutation. For example, since single-strand interruptions occur during recombination, the frequently observed correlation between frameshift mutation and crossing over is explained. The induction of frameshift mutations by ultraviolet irradiation (Drake, 1963*b*, 1966*c*) is explicable as a consequence of breaks introduced by excision type repair processes (Setlow and Carrier, 1966). In the case of acridine mutagenesis of bacteriophage T4, a special factor becomes important. Because the T4 chromosomes are circularly permuted, their ends may fall at any point along the map, whereas in most other organisms, chromosome ends are either fixed or nonexistant. Since the chromosome ends contain long-lasting strand interruptions, Streisinger supposed that most frameshift

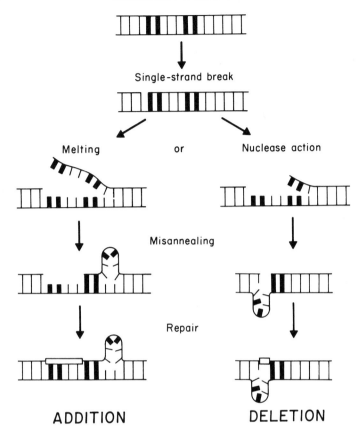

FIGURE 12-8. *Hypothetical mechanism for the production of frameshift mutations by misrepair, after Streisinger et al. (1966). The heavy bars in the schematic DNA molecules represent short regions of base pair homology. The open line segments at the bottom represent newly synthesized DNA. The addition mutant has gained four bases, resulting in a reading-frame shift to the left; the deletion mutant has lost four bases, resulting in a reading-frame shift to the right.*

mutations arise there instead of near recombinational breaks. This assumption therefore explains the apparent association of frameshift mutations with terminal redundancy heterozygotes.

What then is the role assigned to acridines in this process? Streisinger suggested that they act to stabilize improbable mispairing configurations, thus increasing the chances of a "repair" process converting the mispaired configurations into heritable lesions. The special mutagenic abilities of the ICR compounds might be similarly rationalized by supposing that they both alkylate and intercalate, thus localizing an acridine residue directly at the repair site of an alkylated residue. The suppression of mutation observed in T4 at high acridine concentrations, and in yeast during mitosis even at lower acridine concentrations

(Magni, von Borstel, and Sora, 1964), could be attributed to a more general inhibition of repair processes by acridines; just such an effect has been observed upon the system which excises ultraviolet-induced thymine dimers (Witkin, 1963).

The main drawback to the Streisinger model is that it predicts that all intercalating compounds should be mutagenic, though, as we have seen, many are not. Even non-intercalating compounds should induce frameshift mutations if they stabilize double-stranded DNA against thermal denaturation. A number of such compounds (streptomycin, spermine, magnesium), however, are unable to revert frameshift mutations at moderately high concentrations (Lerman, 1964b). (Streptomycin may be mutagenic under some conditions [Fernandez, Haas, and Wyss, 1953], but it is not known if frameshift mutations are induced.) If the degree of stabilization correlated well with mutagenicity among acridines, it is conceivable that some minimal degree of stabilization was required over very short regions in order to preserve misannealed configurations (Figure 12-8). Riva, however, observed no such correlation (Table 12-3). The point which is clear, therefore, is that no clear basis exists for rationalizing the mutagenicities of intercalating compounds. In addition, agents (other than ultraviolet irradiation), such as nitrous acid and alkylating compounds, which probably also produce single-strand breaks are not generally able to induce frameshift mutations, although nitrous acid may produce them in yeasts (Magni, quoted in von Borstel, 1966).

TABLE 12-3

MUTAGENICITY AND DNA STABILIZATION BY ACRIDINES*

Compound	Strong Binding	Weak Binding	Mutagenicity
5-Aminoacridine	100	0.95	strong
5-Amino-10-methylacridinium	100	0.80	none
10-Methylacridinium	67		none
Acridine orange	50		weak

*Data from Riva (1966), Lerman (1964b), and Orgel and Brenner (1961). Strong binding: (increase in DNA melting temperature T_m)/(moles acridine bound per mole of nucleotide). Weak binding: maximum value of (moles acridine bound per mole of nucleotide) with increasing acridine concentration.

The complexity of the rules which govern the mutagenicity of different acridines may be a consequence of the problems which a repair enzyme encounters when it attempts to act upon a region of DNA whose backbone has been elongated and distorted by acridine intercalation. As can be surmised from Figure 12-8, if the mispaired region were to be stabilized by acridine intercalation, the repair enzyme would have to act at almost the same site. The configuration of the broken ends of the DNA molecule which are to be joined, and the fit between the repair enzymes and the nearby portions of the DNA molecule, might then be rather subtly sensitive to the exact nature of the acridine which is present.

13. Chemical mutagenesis

A large number of compounds have been tested for mutagenicity in microbial systems. In this discussion, emphasis will be placed on mutagens whose mechanisms of action are at least moderately well established. A group of compounds which appear to reduce mutation rates will also be considered. Since the base analogues and acridines have been discussed in detail in conjunction with transition and frameshift mutations, little need be said about their mechanisms of action.

BASE ANALOGUES

Litman and Pardee (1956) observed that 5-chlorouracil and 5-iodouracil, as well as 5-bromouracil, were mutagenic. On the other hand, 5-fluorouracil is not mutagenic for bacteriophage T4, nor is the corresponding deoxyribonucleotide (FUDR). FUDR is a potent and irreversible inhibitor of thymidylate synthetase (Cohen et al., 1958), but is usually not incorporated into DNA. However, 5-fluorouracil is readily incorporated into RNA, and is a potent mutagen for RNA viruses such as bacteriophage f_{Can} (Davern, 1964), poliovirus (Cooper, 1964), and TMV (tobacco mosaic virus) (Kramer, Wittmann, and Schuster, 1964). In the case of TMV, the observed amino acid substitutions are explicable on the basis of the transitions $U \rightarrow C$ and $A \rightarrow G$ (Wittmann and Wittmann-Liebold, 1966). Fluorouracil can also induce nonheritable "mutations" in messenger RNA which may result in altered proteins (Champe and Benzer, 1962b; see Chapter 17).

Bromodeoxyuridine (BUDR) is a better mutagen than bromouracil (Freese, 1959b; Litman and Pardee, 1960), presumably because BUDR competes with thymine better than does bromouracil (Laird and Bodmer, 1967). However, inhibition of thymine synthesis by aminopterin, FUDR, or sulfanilamide can force virtually complete substitution of thymine by bromouracil.

While many base analogues have been tested for their effects upon mutation rates, radiosensitivity, and macromolecular syntheses, reports of mutational responses are meager. Scott (1962) reported that 4,5-dihydro-4-methyluracil is a potent mutagen in *Neurospora*, and Gottschling and Heidelberger (1963) reported that 5-trifluoromethyluracil is a potent mutagen in bacteriophage T4.

Freese (1968) reported that 6-hydroxyaminopurine and 2,6-dihydroxy-aminopurine induce both transitions in T4. Halle (1968) reported that 5-azacytidine is strongly mutagenic for RNA-containing arboviruses. A number of purine and pyrimidine analogues which interfere with the synthesis of nucleic acid precursors are also mutagenic, but probably not because of errors of incorporation and replication. These compounds are discussed in a later section.

ACRIDINES

In addition to inducing frameshift mutations, acridines exhibit a number of other mutagenic capabilities about which rather little is known. Dulbecco and Vogt (1958) observed mutagenesis of intracellular polioviruses by proflavin; since frameshift and chain-terminating mutations cannot be recovered when they occur in indispensable genes, and since the minute viruses probably contain exclusively indispensable genes, the poliovirus mutants probably contained base pair substitutions.

Ball and Roper (1966) observed that many of the mutants induced by acridines in *Aspergillus* were unstable and were probably attributable to aneuploidy (changes in the number of chromosomes in a cell, as distinguished from microlesions or macrolesions). This result suggests that the dye may induce aberrant chromosome segregation during cell division.

Acridines are sometimes able to "cure" bacteria of episomes (Hirota, 1960). [Episomes are small genetic elements which may exist in either of two forms, as fragments integrated into the cellular chromosome or as autonomous, self-replicating cytoplasmic entities (Jacob and Wollman, 1961).] Watanabe and Fukasawa (1961) observed that the curing process could be promoted by ultraviolet irradiation, and Clowes and Moody (personal communication) observed a correlation between curing and mutagenesis of bacterial genes. These results suggest that curing may result from lethal mutations of the episome. Kinetic studies of the curing process (Stouthamer, DeHan, and Bulten, 1963), however, suggest that acridines differentially inhibit the replication of episomes, as compared to the replication of bacterial chromosomes, so that the episomes are simply diluted out by cell division. Both curing and episomal mutation have recently been achieved by treating bacteria with sodium dodecyl sulfate (Tomoeda et al., 1968). Since this detergent attacks the cell surface, it may interfere with episomal replication indirectly. Recent observations of Silver, Levine, and Pilelman (1968) suggest that acridines can bind to the surface of the cell as well as to DNA. Acridine curing may therefore occur by a process which is quite distinct from ordinary acridine mutagenesis.

NITROUS ACID

Nitrous acid has been used extensively for inducing mutations in TMV. Wittmann and Wittmann-Liebold (1966) observed that amino acid substitutions induced at 26 different sites were overwhelmingly explicable on the basis of transitions, A → G in 8 cases, and C → U in 17 cases; a pu → py change was required to explain one replacement. An earlier summary (Siegel, 1965) of 63 amino acid replacements indicated that 22 were explicable as A → G transitions, 33 as C → U transitions, and 8 as other changes (A → C four times; G → A two times; and double mutations consisting of C → G and A → U two times). The significance of the less frequent substitutions is obscure because of the possibility of their spontaneous origin. In the *E. coli* tryptophan synthetase A protein, Yanofsky, Ito, and Horn (1966) observed a mixture of A:T → G:C and transversional changes, but the spontaneous backgrounds made very substantial contributions to these replacements. Similar observations were reported by Weigert and Garen (1965), who studied the reversion of UAG mutants of the alkaline phosphatase gene. Malling and de Serres (1968a) concluded that most mutants induced in *Neurospora* contained GC base pairs at the mutant site.

T4rII mutations induced by nitrous acid were induced to revert with base analogues (Freese, 1959c), and most mutations produced by base analogues and by nitrous acid were also induced to revert by nitrous acid (Bautz-Freese and Freese, 1961; Champe and Benzer, 1962b). Several spontaneous mutants (probably frameshift mutants) were not induced to revert by nitrous acid, nor were frameshift mutations in bacteria induced to revert by nitrous acid (see Chapter 12). Tessman, Poddar, and Kumar (1964) concluded that nitrous acid induces all four transitions in bacteriophage S13, but that the T → C transition is induced only weakly.

Tessman (1962b) reported that nitrous acid induces large deletions in bacteriophage T4, and Magni (quoted in von Borstel, 1966) reported that very high concentrations of nitrous acid produce frameshift mutations in yeast. Beckwith, Signer and Epstein (1966) reported that it induced deletions in *E. coli*. Nitrous acid has been employed as a mutagen in many other systems, but the mutations produced have not been closely characterized. However, many are temperature-sensitive or leaky, suggesting the presence of base pair substitutions.

The most striking action of nitrous acid upon nucleic acids is oxidative deamination (Schuster and Schramm, 1958; Schuster, 1960a and b), which produces the transformations guanine → xanthine, adenine → hypoxanthine, and cytosine → uracil. Other as yet uncharacterized reactions also occur (Schuster and Wilhelm, 1963; Freese, 1963; Orgel, 1965). The deamination of cytosine, for example, readily explains the transition C → U (or C → T). Since hypoxanthine should pair well with cytosine (Figure 13-1), the transition A → G is also readily explained. It seemed initially that xanthine should pair with cytosine, and that deamination of guanine would therefore not result in mutations. Xanthine, however, is apparently a "nonsense" base (Richardson, Schildkraut, and Kornberg, 1963; Michelson and Grunberg-Manago, 1964), presumably because at neutral pH it is mostly ionized, even in polynucleotides. As a result, deamination of guanine is likely to be lethal, depending in part upon whether or not the complementary cytosine residue (in DNA) is sufficient to insure gene function.

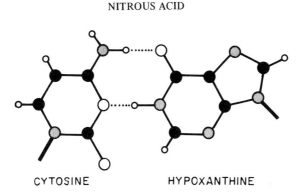

CYTOSINE   HYPOXANTHINE

FIGURE 13-1.   *The cytosine-hypoxanthine base pair in nitrous acid mutagen-esis; symbols are the same as in Figure 1-2.*

Kotaka and Baldwin (1964) studied the replication by DNA polymerase of d(AT) copolymer treated with nitrous acid. In the presence of only dATP and TTP, the priming activity of the template was inactivated. Addition of dCTP partly restored its activity, strongly suggesting that A → G transitions had been induced *in vitro*. Robinson, Tessman, and Gilham (1969) have detected A → G transitions in the RNA of a mutant of bacteriophage MS2 by direct chemical analysis.

The relative rates of deamination as functions of pH are well correlated with the rates of mutation and inactivation in bacteriophage T2 (Schuster, 1960 *a, b*; Schuster and Vielmetter, 1961; Vielmetter and Schuster, 1960 *a, b*). In DNA, guanine reacts much more rapidly than adenine and cytosine (or 5-hydroxymethylcytosine). As the pH is reduced from 5.0 to 4.2, the deamination rate of cytosine and adenine increases 90-fold, whereas the deamination rate of guanine increases 35-fold. At the same time, the mutation rate increases 80-fold, whereas the inactivation rate increases 30-fold.

The kinetics of deamination and mutation sometimes depend rather strongly both upon secondary structure of the nucleic acids, and upon their interactions with protein. When TMV RNA is treated, guanine, cytosine, and adenine all react approximately equally (Schuster and Schramm, 1958), but when the intact virus is treated, guanine residues are strongly protected (Schuster and Wilhelm, 1963). Variable results have been reported when transforming DNA has been treated with nitrous acid in the native and the denatured states. No difference in reaction rates was observed using *B. subtilis* DNA (Strack, Freese, and Freese, 1964), but *Hemophilus* DNA has been reported to be mutable only in the denatured (single-stranded) state, although this was not the case with pneumo-coccal DNA (Luzzati, 1962; Horn and Herriott, 1964).

It was noted above that a few mutations induced in the TMV, T4*r*II, and S13 systems were not explicable as ordinary transitions. There is little chemical evidence to suggest their origin, if indeed they were actually induced and were not simply contaminating spontaneous mutations. Michelson and Monny (1966) have observed that polyxanthylic acid is unable to complex with polycytidylic acid, but that it is able to complex with polyadenylic acid. It is also able to complex with polyuridylic acid, but not with polybromouridylic acid and therefore probably not with polythymidylic acid. Although suggestive of additional mutational effects resulting from the deamination of guanine, these complexes probably involve pairings which are sterically quite improbable in normal DNA. Shapiro and Pohl (1968) report that 2-nitrohypoxanthine is one of the principal side products obtained from guanine, and that small amounts of 8-nitroxanthine also appear; it is not obvious how either of these compounds could engage in mutagenic mispairing, however.

A very different type of reaction between DNA and nitrous acid has been described by Becker, Zimmerman, and Geiduschek (1964). At pH 4.15, one covalent cross-link was formed for each four deaminations. It is appealing to suppose that cross-linking is the reaction which produces the deletions observed by Tessman (1962), but no experimental connection between the two has yet been reported.

HYDROXYLAMINE

Evidence was discussed in Chapter 10 indicating that hydroxylamine reacts mainly with cytosine and guanine, and that it induces the transition $C \rightarrow T$ with very high specificity (so long as virus particles or free nucleic acids are treated). The chemistry of the reaction between hydroxylamine and cytosine, and the corresponding theory of mutagenesis, have been recently reviewed in depth by Phillips and Brown (1967). Although the nature of the mutagenically significant reaction was in doubt for some time, a number of studies now implicate the two pathways shown in Figure 13-2. In both cases, the tautomeric form on the right should be capable of efficient pairing with adenine (see Figure 10-1). In fact, the efficiency of mispairing estimated in an *in vitro* system, which is discussed below (Phillips, Brown, and Grossman, 1966), approaches unity. Hydroxylamine is an unusual mutagen in that it appears to act on cytosine and 5-hydroxymethyl-cytosine (5HMC) in different ways (Janion and Sugar, 1965 *a, b*), nevertheless achieving a common mutational specificity. It is still possible, however, that a portion of the reaction with cytosine in DNA procedes in a manner analogous to the reaction shown for 5HMC, producing N^6-hydroxycytosine (Lawley, 1967).

The reactivity of cytosine depends strongly upon secondary structure. T4 is a thousand times more mutable than is *B. subtilis* transforming DNA, and denatured *B. subtilis* DNA is even more mutable than is T4 (Freese and Strack, 1962). The reactivity of T4 probably reflects the fact that the DNA inside the head of bacteriophages is partially denatured (Tikchonenko et al., 1966; Gorin et al., 1967; Maestre and Tinoco, 1967). Ethylene glycol, which strongly reduces the amount of secondary structure in DNA, also promotes hydroxylamine mutagenesis of transforming DNA (Freese and Strack, 1962). The reactivity of cytosine also depends upon substitutions at the 5 position. Thus 5HMC is some ten times less reactive than is cytosine (Freese, Bautz-Freese, and Bautz, 1961).

4,5-DIHYDRO-4-HYDROXYLAMINOCYTOSINE

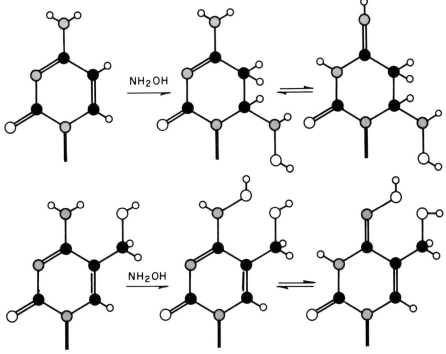

N⁶-HYDROXY-5-HYDROXYMETHYLCYTOSINE

FIGURE 13-2. *The main reactions of hydroxylamine with cytosine (top) and with 5HMC (bottom), together with the presumed mutagenic tautomers of the reacted bases (right).*

The mutagenicity of hydroxylamine has been studied in an *in vitro* system which employs the *Micrococcus lysodeikticus* RNA polymerase (Phillips, Brown, Adman, and Grossman, 1965). This enzyme is capable of utilizing single-stranded RNA as a template for the synthesis of the complementary polyribonucleotide; for instance, polycytidylic acid directs the synthesis of polyguanylic acid from GTP. Hydroxylamine treatment of the polycytidylic acid template inactivates its priming ability, but the addition of ATP restores at least part of the lost activity, the adenine residues being incorporated into the product. The probability that a monoreacted cytosine residue will direct the incorporation of an adenine rather than a guanine residue is close to unity (Phillips, Brown, and Grossman, 1966). The kinetics of the reaction strongly indicate that the monoreacted cytosine product (4,5-dihydro-4-hydroxylaminocytosine, Figure 13-2) is the mutagenic species; both it and N⁶-hydroxycytosine can be readily converted by hydroxylamine into N⁶-hydroxy-4,5-dihydro-4-hydroxylaminocytosine, which appears to be lethal in the RNA polymerase system in the sense that it irreversibly blocks polymerization. N⁶-hydroxycytosine (produced synthetically, and then used to construct a template polymer) has also been tested in the RNA polymerase system (Brown, Banks, and Grossman, personal communication). It too behaves like uracil (Brown, Hewlins, and Schell, 1968), although its mispairing efficiency is only about half that of the presumed 4,5-dihydro-4-hydroxylaminocytosine species. The experiments with the RNA polymerase system thus appear to fully support the two schemes illustrated in Figure 13-2. However, the obvious possibility still exists that the RNA system behaves differently from normal DNA replication.

Under certain conditions hydroxylamine is observed to produce a high frequency of lethal hits. This reaction can be suppressed by using high concentrations of hydroxylamine itself, by the removal of oxygen, or by the addition of compounds such as catalase, peroxidase, pyrophosphate, or EDTA (Freese, Bautz-Freese, and Bautz, 1961; Freese and Freese, 1965; Van der Pol and Van Arkel, 1965; Freese, Freese, and Graham, 1966; Tessman, 1968). This mode of inactivation, therefore, very probably procedes by the formation of peroxides and free radicals.

ALKYLATING AGENTS

The first chemical mutagen to be discovered was mustard gas, an alkylating agent (Auerbach and Robson, 1947); since that time, a large number of alkylating agents have been demonstrated to be mutagenic for a variety of organisms, from the most simple viruses to mammals and higher plants. A very useful monograph on the genetic effects of alkylating agents has recently been prepared by Loveless (1966), who also discovered the mutagenic effects of alklyating agents upon bacteriophages (Loveless, 1958). In addition, Lawley (1966) has prepared a useful review of the chemistry of alkylation of nucleic acids. A number of geneticists have become increasingly conscious in recent years of the great potential danger of alkylating agents as polutants of man's environment. Many of these agents are among the most potent mutagens, but exhibit relatively low toxicities and tend to produce delayed genetic effects. The practical dangers involved in the production of these compounds, many of which are important in the chemical industries, arise from two possibilities: accidental release of large quantities upwind or upstream from sizable centers of population, and undetected mutagenicity for years and possibly decades afterwards.

Only a few alkylating agents will be discussed here, namely, those which have been employed in studies of mutational mechanisms. Their names, structures, and abbreviations appear in Figure 13-3. (The compounds pictured in Figure 13-4 are also presumably alkylating agents.)

The mutational specificities of several alkylating agents have been determined in bacteriophage systems. In T4, for example, EMS and EES very strongly induce the reversion of those mutants which are also induced to revert by hydroxylamine (Krieg, 1963b; Bautz and Freese, 1960; Freese, 1961). These agents, therefore, strongly induce G:C $\rightarrow$ A:T transitions. EES and EMS also induce the reversion of two other types of mutants, namely, those both induced and reverted by base analogues but *not* by hydroxylamine, and those induced by

CHEMICAL NAME	COMMON NAME	STRUCTURE
DI – (2 – CHLOROETHYL) SULFIDE	MUSTARD GAS	$Cl-CH_2-CH_2-S-CH_2-CH_2-Cl$
ETHYLMETHANE SULFONATE	EMS	$CH_3-CH_2-O-SO_2-CH_3$
ETHYLETHANE SULFONATE	EES	$CH_3-CH_2-O-SO_2-CH_2-CH_3$
DIETHYLSULFATE	DES	$CH_3-CH_2-O-SO_2-O-CH_2-CH_3$
N – METHYL – N' – NITRO – N – NITROSOGUANIDINE	NG	$HN = C - NH - NO_2$ $O=N-N-CH_3$

FIGURE 13-3.   *Some frequently used alkylating agents.*

proflavin. The first type of mutant consists of transitions containing A:T base pairs at the mutated sites, and the second type of mutant contains frameshift lesions. The responses of these mutants are quite weak, but the average response of the A:T mutants are slightly stronger (about 4- to 10-fold over the spontaneous background) than the average response of the frameshift mutants (about 2-fold over the spontaneous background). Most of these responses are so weak that they might even be interpreted as arising indirectly: since alkylation lesions are subject to excision-type repair in many organisms (see Loveless, 1966, and Lawley, 1966), both frameshift mutations and mutations involving A:T sites may occur spontaneously during the resynthesis of the excised regions. A prediction of this model is that if the appropriate nonexcising mutants could be obtained, then the weak mutational responses would no longer be observed.

Freese (1961) detected a class of mutants which could be induced to revert by EES, but which were not reverted at all by base analogues. The reasonable assumption was made that these mutants contained transversions (though the actual pathways have not yet been worked out). Transversions, however, appear to be decidedly less frequent than transitions.

The mutations induced in T4 by EMS are characterized by a pattern of delayed appearance (Green and Krieg, 1961; Krieg, 1963b). Single-burst analyses of newly arisen mutant clones indicate that they arise with approximately equal probability throughout the period of DNA replication. (This pattern of delayed mutation should not be confused with the patterns of delayed expression which were outlined in Chapter 10.)

An analysis of the mutations induced in bacteriophage S13 by EMS has largely confirmed the results obtained from the T4 system (Tessman, Poddar, and Kumar, 1964). Both the G → A and the C → T transitions were strongly induced. The T → C transition was regularly but weakly induced. The A → G transition was rarely or never induced.

When *E. coli* tryptophan synthetase mutants were induced to revert with EMS, a much wider range of base pair substitutions was deduced from the resulting amino acid replacements than had been detected in viral systems (Yanofsky, Ito, and Horn, 1966). The transition G:C → A:T was not observed at all, although it probably simply escaped detection. The transition A:T → G:C was strongly induced. The transversions A:T → T:A and G:C → C:G were also observed. In tobacco mosaic virus, the transitions A → G and C → U have both been observed (Funatsu and Fraenkel-Conrat, 1964). In *Neurospora crassa,* most of the mutants induced by EMS were transitions, the A:T → G:C pathway being favored; however, transversions, frameshift mutations, and probably even deletions were also induced (Malling and de Serres, 1968*b*).

The major chemical reactions between alkylating agents and nucleotides occur at the N7 position of guanine, at the N1 and N3 positions of adenine, and at the N1 positions of cytosine and thymine (see Lawley, 1966). The phosphates may also react, but this reaction is unlikely to be of importance in the strong mutational processes since it should be almost completely nonspecific. Analyses of a variety of types of DNA treated with EMS and MMS reveal that alkylation occurs mainly at the 7 position of guanine (79—92 percent), frequently at the 3 position of adenine (8—19 percent), and occasionally or rarely at the 1 position of adenine (2—3 percent) and the 1 position of cytosine (trace). Alkylated thymine was not produced at detectable levels. Similar results were obtained using either the T-even bacteriophages or their free DNA. In studies with nitrogen mustard, only guanine was detectably reacted (at the 7 position) when double-stranded DNA was treated with moderate doses; guanine residues in DNA reacted 50-fold faster than did guanine riboside (Price et al., 1968).

Loveless (1969) has very recently reported that deoxyguanosine is alkylated at the O-6 position at a very significant rate.

One of the guanine reactions is likely to produce the G → A transitions. Lawley and Brookes (1961) pointed out that the N-7 alkylation product readily ionizes, and might then pair with thymine (see Figure 10-2). The free N-7 alkylated species is not mutagenic in T4 (Rhaese, 1968), but may never be converted to the immediate precursor of DNA synthesis. During *in vitro* DNA synthesis, a pH high enough to extensively ionize guanine does not induce a high mispairing rate (Trautner et al., 1962). Furthermore, methyl methanesulfonate efficiently alkylates the N-7 position, but is not mutagenic for the T-even bacteriophages. The O-6 alkylation product, on the other hand, resembles the enol form of guanine (see Figure 10-1), and might pair well with thymine. The pattern of delayed mutation observed by Green and Krieg (1961), however, suggests that mispairing occurs with a probability considerably less than unity.

It is not clear how the other alkylated bases would engage in mutagenic mispairing, although the data of Tessman, Poddar, and Kumar (1964) suggest that the alkylated pyrimidines do mispair moderately often. Alkylation of the pyrimidine and adenine ring nitrogens which are involved in base pairing should interfere drastically with pairing of any sort. The possibility should therefore be considered that these alkylated bases block hydrogen bonding more or less completely, but still allow the insertion of a purine opposite an alkylated pyrimidine (or a pyrimidine opposite an alkylated adenine) with an increased probability of an erroneous insertion.

Ludlum and Wilhelm (1968) studied RNA synthesis performed by *M. lysodeikticus* RNA polymerase using a methylated polycytidylic acid template, and observed G → U transversions. The relevance of this substitution to DNA synthesis is unclear, but it is obviously of great potential interest.

Depurination frequently results from the alkylation of guanine and adenine, and has often been suggested to be the source of mutations, especially of transversions (Bautz and Freese, 1960; Freese, 1961). A number of arguments were put forth against this proposal in Chapter 11. The most impressive of these is that depurination continues long after the treatment of DNA with alkylating agents, whereas only lethal, but not mutational hits accrue during the same period. In addition, ethylation is much more strongly mutagenic for T-even bacteriophages than is methylation, whereas the methylated bases depurinate more easily. Depurination may not even be lethal much of the time, since Brookes and Lawley (1963) observed that about one hundred alkylated guanine residues hydrolyze per lethal hit. Perhaps only those depurinations which cannot or happen not to be repaired become lethal, as is often the case with ultraviolet-induced lesions.

MISCELLANEOUS CHEMICAL MUTAGENS

Here our discussion deals with a number of compounds whose mutational specificities are usually only poorly understood, and whose mechanisms of action are even less well characterized. They are included because they offer some promise of exhibiting interesting specificities or mechanisms, or simply because they are good mutagens in bacterial or viral systems.

One of the most potent mutagens described to date for microorganisms is Adelberg, Mandel, and Chen, 1965). In bacteria it induces both transitions and transversions, but not frameshift mutations (Eisenstark, Eisenstark, and van Sickle, 1965; Whitfield, Martin, and Ames, 1966). It may, however, induce large deletions (Langridge and Campbell, 1969). Cerdá-Olmedo, Hanawalt, and Guerola (1968) demonstrated a spectacular correlation in synchronized cultures of *E. coli* between the distance of a marker from the origin of DNA replication, and its period of maximum revertibility by NG, indicating that the replication point was preferentially mutagenized. Although NG mutagenesis of exponentially replicating cells produces a random distribution of mutations, mutagenesis of stationary phase cells produces mutations strongly clustered in one region of the chromosome (Botstein and Jones, 1969). Several workers have observed that NG sometimes induces tight clusters of mutations, a circumstance which can easily confuse the analysis of NG-induced mutants. In bacteriophages, NG is only effective when applied to infected cells (Goldfarb, Nesterova, and Kuznetsova, 1966; Zampieri, Greenberg, and Warren, 1968); it produces all possible transitions in S13 and T4 (Baker and Tessman, 1968). In tobacco moasic virus, however, NG is strongly mutagenic when applied to free virus particles, and paradoxically only weakly mutagenic when applied to naked viral RNA (Singer and Fraenkel-Conrat, 1967); both C → U and A → G transitions are induced.

NG is an alkylating agent; both 7-methylguanine and 3-methyladenine are produced in DNA, although pyrimidine derivatives have not been detected (Lawley, 1968). However, NG also tends to decompose into a variety of compounds, some of which, such as diazomethane, might be more mutagenic than the parent compound (Cerdá-Olmedo and Hanawalt, 1967, 1968; Singer and Fraenkel-Conrat, 1967). Furthermore, a definite anticorrelation appears to exist between alkylation and NG mutagenesis. In TMV, methylation of guanine and cytosine residues is sharply depressed when the RNA is treated in formamide (compared to water), whereas mutagenesis is considerably increased (Singer et al., 1968). Comparable *in vivo* experiments in *E. coli* also tend to separate the two processes (Süsmuth and Lingens, 1968). The mechanism of action of NG therefore remains an intriguing mystery.

Streptomycin at very inhibitory concentrations induces base pair substitution (host range) mutations in T2 (Fernandez, Haas, and Wyss, 1953), but does not induce frameshift mutations (Lerman, 1964*b*). Although it readily complexes with DNA, streptomycin also induces errors in translation (see Gorini and Beckwith, 1966), and may therefore act somewhat indirectly as a mutagen.

Prell (personal communication) observed that stocks of bacteriophage P22 stored aerobically over chloroform accumulate both mutations and lethal damages (Prell, 1960 *a, b*). Chloroform is commonly used to sterilize bacteriophage stocks. This effect has not been observed in T4.

Because of the free-end phenomenon postulated for the formation of frameshift mutations (see Chapter 12), DNA fragments might be expected to be mutagenic. Demerec (1962, and personal communication) observed that many mutants of *Salmonella* could be weakly reverted by "transduction" with a bacteriophage grown on the same strain, or even when the donor strain contained a deletion of the region which was mutated in the acceptor strain. Both Yoshikawa (1966) and Chernik and Krivisky (1968) observed that transformation of *B. subtilis* with DNA from the same strain produced a very high mutant yield, up to 0.2 auxotrophs per *competent* cell. These mutations were certainly not very often frameshift lesions, however, since about a third of them were leaky, and many were temperature-sensitive.

Mn^{++} is a powerful mutagen for *E. coli* (Demerec and Hanson, 1951; Böhme, 1961*a*) and for replicating bacteriophage T4 (Orgel and Orgel, 1965). Nearly all of the induced T4*r*II mutations are revertible with base analogues, indicating that transitions are induced. Mn^{++} is known to alter the properties of DNA polymerase, since it induces the enzyme to accept ribonucleoside triphosphates *in vitro* (Berg, Fancher, and Chamberlin, 1962).

A long delay in filling a position during DNA replication, brought about by a shortage of the appropriate precursor residue, might favor the incorporation of incorrect nucleotides. A number of compounds which interfere with the formation of the normal pyrimidine and purine precursors in *E. coli* are also mutagenic. The inhibitors of purine production include azaserine (Iyer and Szybalski, 1958); benzimidazole, caffeine, paraxanthine, tetramethyl uric acid, theobromine, and theophylline (Novick, 1956); and 6-mercaptopurine and 5-nitroquinoxaline (Greer, 1958). The inhibitors of thymine production include 5-aminopurine (Greer, 1958), and, of course, thymine deficiency itself (Kanazir, 1952; Coughlin and Adelberg, 1956). Thymine starvation appears mostly to produce A:T→G:C transitions, probably by increasing the probability of inserting cytosine into a thymine position (Pauling, 1968); however, transversions and other types of lesions are also induced (Holmes and Eisenstark, 1968). It should be noted, however, that some of these agents may induce mutations by other means, for instance, azaserine by alkylation (see Loveless, 1966), and caffeine and its relatives by interfering with the repair of premutational lesions. For instance, Bridges, Law, and Munson (1968) reported that thymineless mutagenesis is abolished in certain *E. coli* mutants deficient in DNA repair.

A number of polyimines originally considered for use as insect chemo-sterilants (Figure 13-4) are quite mutagenic for bacteriophage T4 (Drake 1963a), for *E. coli* (Szybalski, 1958), and for *Neurospora* (Kaney and Atwood, 1964). These compounds are likely to function as alkylating agents. In T4, extrapolation from the rate of production of *r* mutations indicates that they produce about 0.02–0.05 mutational hits per lethal hit, except for Apholate, which may not penetrate well to the viral DNA.

Several compounds related to hydroxylamine are mutagenic, including N-methylhydroxylamine (CH_3NHOH), which mutates transforming DNA but not bacteriophage T4, and methoxyamine (NH_2OCH_3), which mutates both (Freese, Bautz-Freese, and Bautz, 1961; Freese and Freese, 1964). Hydrazine (NH_2NH_2) is mutagenic for bacteria and bacteriophages, and is reputed to induce transitions in T4 (Freese, Bautz, and Freese, 1961; Orgel, quoted in Phillips and Brown, 1967). Unlike hydroxylamine, however, it is very lethal, and is most highly mutagenic at an alkaline pH. Brown, McNaught, and Schell (1966) suggest that hydrazine reacts with cytosine to produce 4,5-dihydrocytosine, which by analogy with 4,5-dihydro-4-hydroxylaminocytosine (Figure 13-2) should behave very much like thymine.

Acid, specifically pH 4.2–5.0 acetate buffer at 37°–45°, is mutagenic for transforming DNA and for bacteriophage T4 (Freese, 1959c; Freese, 1961; Strack, Freese, and Freese, 1964). Most acid-induced *r*II mutants are revertible by base analogues. Acid treatment is capable of inducing revertants from G:C transitions, but not from A:T transitions; it is also capable of inducing revertants of those transversions which respond to alkylating agents (probably containing G:C base pairs). Shapiro and Klein (1966) report the acid-induced deamination of cytosine to uracil, which might explain the G:C → A:T transitions. The Freese group has postulated that the transversions arise from depurination, but this seems unlikely for the reasons discussed previously.

Certain acridines (such as acriflavin) and methylated purines (such as

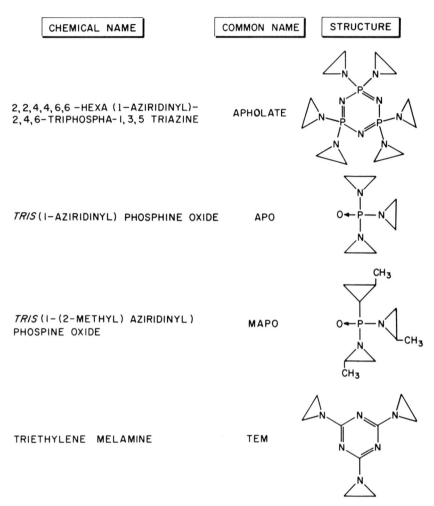

CHEMICAL NAME	COMMON NAME	STRUCTURE

2,2,4,4,6,6 –HEXA (1–AZIRIDINYL)– 2,4,6-TRIPHOSPHA-1,3,5 TRIAZINE — APHOLATE

TRIS (1–AZIRIDINYL) PHOSPHINE OXIDE — APO

TRIS (1–(2-METHYL) AZIRIDINYL) PHOSPINE OXIDE — MAPO

TRIETHYLENE MELAMINE — TEM

FIGURE 13-4.   *Some mutagenic chemosterilants and the aklylating agent TEM.*

caffeine) have been reported to interfere with the repair of lesions induced by ultraviolet irradiation, to be moderately mutagenic themselves, and to act synergistically with ultraviolet mutagenesis (Novick and Szilard, 1951; Lieb, 1961; Witkin, 1961; Doneson and Shankel, 1964; see also Chapters 12 and 14). On the assumption that repair processes frequently remove spontaneous premutational lesions, these compounds may therefore exert an "antiantimutagenic" action. Shankel and Kleinberg (1967) have presented evidence which suggests that caffeine binds to a repair enzyme (although it certainly also interacts strongly with DNA), whereas acriflavin binds to an ultraviolet-induced lesion. Caffeine is not a mutagen in bacteriophage T4, however, nor does it promote ultraviolet mutagenesis (Drake, 1964).

ANTIMUTAGENS

Reports have appeared from time to time indicating that certain compounds can reduce the rate, not only of induced, but also of spontaneous mutation. Spontaneous mutation rates are generally reduced by small factors. The most curious result is that some of these agents are antimutagenic under some conditions, but are mutagenic under other conditions. Novick (1956) observed that spontaneous mutation rates of replicating *E. coli* were reduced about 3-fold by adenosine, guanosine and inosine, and that caffeine mutagenesis was also inhibited. Webb and Kubitschek (1963) observed that acridine orange behaved identically under similar conditions, and Magni, von Borstel, and Sora (1964) and Puglisi (1967) reported that 5-aminoacridine was antimutagenic for mitotically multiplying yeasts. Puglisi (1968) also reported that both actino-mycin D and basic fuchsine are strongly antimutagenic in *S. cerevisiae*. Grigg and Stuckey (1966) observed that even caffeine could act as an antimutagen if it was applied to stationary phase cultures of *E. coli*. Johnson and Bach (1965, 1966) reported that spermine was an effective antimutagen for multiplying *E. coli* and *Staphlococcus aureus*, reducing the frequency not only of spontaneous mutation, but also of mutation induced by caffeine, ultraviolet irradiation, 2-aminopurine, and a mutator gene; the acridine quinacrine was also observed to be antimutagenic. In bacteriophage T4, however, caffeine and spermine are neither mutagenic nor antimutagenic, nor have they been observed to interfere with repair processes.

There is little basis upon which to build an adequate explanation of antimutagenicity. Although the interaction of some of these compounds with repair systems, as well as their occasional mutagenicity, can be interpreted as expressions of two or more quite different interactions with the genome and its associated replication apparatus, an imaginative reader may wish to search for a single underlying mechanism. Since these effects are not observed in bacterio-phage T4, even though the compounds themselves interact with viral DNA in much the same way that they interact with cellular DNA, it may be more profitable to concentrate upon their interactions with the enzymes which operate upon DNA.

14. Radiation mutagenesis

During the first three decades following the discovery of X-ray mutagenesis in *Drosophila* (Muller, 1927), mutations were routinely obtained in many organisms by the use of radiations. Even now, however, only a bare beginning has been made in understanding the mechanism of radiation mutagenesis.

Depending upon their characteristic wavelengths, different radiations produce quite different types of mutational lesions, and we shall consider separately the production of mutations by X-rays, ultraviolet light, visible light, and finally heat, which is somewhat arbitrarily included among the radiations.

IONIZING RADIATIONS

A huge literature has accumulated concerning the mutagenic effects of ionizing radiations on eukaryotes. (Eucaryotic organisms such as fungi and man possess complex chromosomes whose DNA is complexed with strongly basic proteins and RNA, whereas procaryotic organisms such as viruses and bacteria possess chromosomes which consist exclusively of DNA.) Many of these mutations appear to be macrolesions which arise from chromosome breaks. The underlying mechanisms are poorly understood, primarily because of insufficient knowledge about the molecular structure of these complicated chromosomes. Mutation induction sometimes exhibits single-hit, and sometimes two-hit kinetics. Webber and de Serres (1965) have shown that those mutations in *Neurospora* which are highly localized (and may be true microlesions) arise with single-hit kinetics, whereas macrolesions arise with two-hit kinetics. Single-hit events necessarily predominate at relatively low doses, whereas two-hit events predominate at high doses. The nature of the mutations which are recovered depends upon the ploidy of the organism. (Ploidy refers to the number of complete sets of chromosomes which an organism possesses. Since most higher animals are diploid, a recessive mutation in one of their chromosomes is phenotypically masked by a wild type allele in the homologous chromosome.)

Numerous reports also exist concerning ionizing radiation mutagenesis in bacteria; probably both microlesions and macrolesions are produced. This literature will not be discussed, however, because it does not seem to offer insights into molecular mechanisms. The inherent advantages of viruses for such studies will by now be clear to most readers. It is only recently, however, that reliable reports have begun to appear concerning ionizing radiation mutagenesis in viral systems. These are summarized in Table 14-1. The irradiation was in each

TABLE 14-1

IONIZING RADIATION MUTAGENESIS OF BACTERIOPHAGES*

Bacteriophage	Mutation	Kinetics	Reference
Kappa	$c^+ \rightarrow c$	one-hit	Kaplan, Winkler, and Wolf-Ellmauer (1960)
T2	$h^+ \rightarrow h$	one-hit	Ardashnikov, Soyfer, and Goldfarb (1964)
ϕX174	$h^+ \rightarrow h$	one-hit	Van der Ent, Blok, and Linckens (1965)
T4	$r \rightarrow r^+$	two-hit	Brown (1966a and b)

* Only examples reinforced by supporting data are cited. The T2 experiments were performed using gamma rays from a Co^{60} source, while the remainder were performed using X-rays of various energies. A weak induction of T4r mutations was also reported by Brown.

case performed upon virus particles suspended in broth. In at least two cases, control experiments were reported demonstrating that irradiation of the broth itself was not mutagenic for the virus particles (Brown, 1966a; Van der Ent, Blok, and Linckens, 1965).

The effects of ionizing radiations upon organisms can be separated into two categories, usually termed direct and indirect (Watson, 1950, 1952). When viruses are irradiated in broth, or in buffer containing protective compounds, only direct effects are observed. When they are irradiated in buffer, killing is much more severe and may continue after irradiation; the differences are attributed to indirect effects. The importance of indirect effects in the production of microlesions is emphasized by the observation that the sexual transfer of DNA into a previously irradiated cell can result in mutagenesis of the transferred DNA (Kada and Marcovich, 1963). The meager genetic results now available do not, however, justify an extended discussion of the radiation chemistry of nucleic acids.

ULTRAVIOLET RADIATION

Ultraviolet mutagenesis has been employed in a great variety of microorganisms for the routine production of mutations. ("Ultraviolet" in this context usually means the 253.7 nm mercury emission obtained from a low-pressure germicidal lamp.) The mechanisms involved have been studied in comparatively few organisms, among which E. coli and T4 predominate. A great deal of attention has been given to the role of repair processes in the fate of primary photochemical lesions, and while many interesting observations have resulted, simple answers certainly have not. It seemed for a few years after the discovery of pyrimidine dimers, a major class of photoproducts, that the photochemistry of the nucleic acids might turn out to be a relatively simple affair, but this is no longer the case. In addition, both viruses and cells have evolved complicated repair systems which often differ among different organisms. Therefore, while much attention has been given to the effects of repair processes upon mutational lesions and photochemical lesions, neither is very well understood. In the discussion that follows, viral systems will be considered first, since some fairly clear results have been obtained with them. The modifying effects of repair systems will be considered, followed by the chemistry of ultraviolet-induced lesions and their repair. Finally, explicit mechanisms for ultraviolet mutagenesis will be suggested.

Many viruses are susceptible to ultraviolet mutagenesis, but simultaneous or near-simultaneous irradiation of the host cell is frequently required. This is particularly true of the temperate and semitemperate bacteriophages. (The temperate bacteriophages such as λ, kappa, and P2 are capable of establishing lysogeny; the semitemperate ones such as T1, T3, and S13 do not establish lysogeny, but do not interfere with host cell metabolism as extensively as do purely lytic bacteriophages.) Irradiation of both cell and virus has been reported to be required for mutagenesis of λ (Weigle, 1953), P2 (Bertani, 1960), T1 (Tessman, 1956), T3 (Weigle and Dulbecco, 1953), and S13 (Tessman and Ozaki, 1960). Jacob (1954) reported that irradiation of the host cell alone induced the v (virulent) mutation in λ, but Weigle (1953) was unable to detect mutations in λ unless the virus was also irradiated; similar observations were made for T1 and S13. Devoret (1965) observed that the induction of v mutations by irradiation of only the host cell required the activity of a cellular repair system (*hcr*), and suggested that damaged bases are incorporated into viral DNA.

When an ultraviolet-irradiated temperate or semitemperate virus is grown in an irradiated host cell, it experiences an *increased* probability of survival compared to growth on unirradiated cells. This phenomenon is called ultraviolet restoration (Weigle, 1953), and is well correlated with mutagenesis in systems in which irradiation of the host cell is also required. Restoration can also be achieved by X-irradiation or alkylation of the cells, or even by coinfecting an unirriadated cell with several heavily irradiated particles, in addition to the lightly irradiated particle which is to be restored (Harm, 1963*b*). The correlation between mutagenesis and ultraviolet restoration is paradoxical, since a decrease in lethality is associated with an increase in the mutation rate. Repair systems may be imagined to act in either of two ways in these systems: they may convert lethal hits into mutational hits, or they may preferentially repair lethal hits rather than mutational hits. The qualitatively different behavior of repair systems in irradiated cells compared to unirradiated cells could result from their *de novo* induction, or from their fractionation by competition between lesions on cellular DNA and lesions on viral DNA. To suggest a specific example of how the process might work, nucleases have been detected in bacteria which specifically degrade irradiated DNA unless repair occurs sufficiently rapidly (Kellenberger, Arber, and Kellenberger, 1959; Strauss, 1962b; Harm, 1963b). If these nucleases recognized potentially mutational lesions in viral DNA, lethality would result; if they were complexed by irradiated cellular DNA, then both survival and mutation of viral genomes would increase. It seems very likely that repair processes, whatever their nature, would vary in efficiency among different cell strains, and Bertani (1960) has reported that restoration and mutagenesis are observed in some strains but not in others.

A further complication may also exist in these systems: competition between repairable lesions in cellular DNA and in viral DNA. In a model system, Harm and Hillebrandt (1963) observed that coinfecting cells with lightly and heavily irradiated T4 depressed the photoreactivation of the lightly irradiated particles, but that this depression could be reversed by prior maximal photoreactivation of the heavily irradiated particles. Note, however, that this type of competition will generally be antagonistic to ultraviolet restoration.

The T-even bacteriophages do not exhibit a requirement for irradiation of the host cell in order for mutations to be induced by ultraviolet irradiation, nor is the survival of irradiated virus affected by known cellular repair processes other than photoreactivation (Dulbecco, 1950; Ellison, Feiner, and Hill, 1960; Harm, 1963a). The glucosylation of the 5HMC residues probably interferes with the ability of most enzymes to act upon T-even DNA. Although some early reports suggested that T-even bacteriophages are mutagenized only during intracellular replication (Latarjet, 1949, 1954), the results were complicated by the still undiscovered phenomenon of phenotypic mixing: h mutations induced in free virus particles do not score until passaged once through a permissive host. Some later workers also experienced difficulties in scoring plaque morphology mutants because of effects upon the first cycle of infection on the plate produced by nonlethal, nonheritable ultraviolet damages. However, it was later shown that free virus particles are mutagenized efficiently (Krieg, 1959; Folsome and Levin, 1961; Folsome, 1962; Drake, 1966c). T4rII mutations are produced with similar mutational specificities, whether induced in free virus particles or during intracellular replication (Drake, 1963b, 1966c).

The mutational specificity of ultraviolet irradiation has been most closely examined in bacteriophage T4 (Drake, 1963b, 1966 c, d). When free virus particles are irradiated, mutations are produced with single-hit kinetics (Krieg, 1959, Folsome and Levin, 1961; Setlow, 1962; Drake, 1966c). Deletions are produced rarely if at all. About half of the induced rII mutations are reverted by proflavin, indicating that they contain frameshift lesions. Nearly all of the remainder are reverted by base analogues, indicating that they contain base pair substitutions. Very few of these are reverted by hydroxylamine. This pattern indicates that most of the mutations arise as $G:C \rightarrow A:T$ transitions. However, a few transversions may also be induced (Brenner and Shulman, personal communications); since many rII transversions are reverted by base analogues, they are not easily detected. The mutational specificity of ultraviolet irradiation has also been studied in bacteriophage S13 (Howard and Tessman, 1964*b*). The mutants were induced under conditions of ultraviolet restoration. One-half to two-thirds of them appeared to contain C → T transitions; the remainder were either T → C transitions, or transversions. As noted previously, frameshift lesions would probably not be recoverable in S13. In *E. coli*, G:C → A:T transitions were readily induced (Yanofsky, Ito, and Horn, 1966), but only about 9 percent of UV-induced mutants contained frameshift mutations (Berger, Brammar, and Yanofsky, 1968). None of four frameshift mutations were induced to revert by UV, but selection against revertants is suggested by the data; both frameshift mutations tested in T4 were induced to revert by UV (Drake, 1963*b*).

Folsome (1962) also studied ultraviolet-induced T4*r* mutants. These mutants, however, bahaved quite differently from those described above. A moderate frequency (8 percent) of deletions appeared, and base pair substitutions accounted for only 11 percent of the *r*II mutants. Of the 14 largest hot spots observed by Drake in the mutational spectrum, only a few were represented in this collection. On the whole, fewer differences were seen by Folsome than by Drake between spontaneous and ultraviolet-induced mutants. The significance of these differences is yet to be determined.

Although insensitive to most types of repair carried out by cellular enzymes, the T-even bacteriophages are susceptible to photoreactivation: if irradiated particles are adsorbed to cells and the complexes are irradiated with intense white light, a considerable fraction of the lethal hits may be repaired even before any significant amount of intracellular viral development occurs. Since at least two types of mutational lesions are induced in T4 by ultraviolet irradiation, it is desirable to know whether either or both are subject to photoreversal. Using a T4 mutant particularly sensitive to photoreactivation, Drake (1966*d*) observed that *both* types of lesion are approximately equally photoreversible.

A bacteriophage system which appears somewhat intermediate between the temperate and the T-even types has been studied by Kaplan and his associates (Ellmauer and Kaplan, 1959; Kaplan, Winkler, and Wolf-Ellmauer, 1960; Winkler, 1963; Steiger and Kaplan, 1964 Winkler, 1965 *a*, *b*; Kaplan, 1966). Bacteriophage kappa, active on strains of *Serratia marcescens*, can be mutagenized without simultaneous irradiation of the host cell. Very high frequencies of *c* mutations are produced, but other mutations which readily appear after chemical mutagenesis are not efficiently produced by ultraviolet irradiation. Unlike T4, where *r* mutations are produced with single-hit kinetics, *c* mutations are produced with two-hit kinetics. A number of treatments affect the final mutant frequency, including white light irradiation of the particles themselves, conventional photoreactivation, and mild heating. Ultraviolet restoration, while not obligatory for mutagenesis, nevertheless stimulates the final mutant frequency two- to four-fold. Complicated responses to chemical or mutational depletion of the ability of the host cell to carry out various types of repair are observed. The *c* mutations fail to appear when the irradiated particles are plated on certain cell strains, even though ultraviolet restoration may still be present.

One possible complication in these studies, consistent with the very high level of mutagenesis, the multiple-hit kinetics of induction, and the dependence on specific host cell strains for the appearance of c mutations, is that the mutations are not actually induced *de novo*, but are introduced by genetic recombination between the infecting virus and a fragment of virus genome present in certain cell strains. This possibility constitutes a general problem when temperate viruses are studied, and will be discussed in Chapter 19.

The analysis of ultraviolet mutational specificities in purely cellular systems has been approached, not by analysis of the chemically induced reversion of mutants, but by comparisons of the repair of mutational lesions with the repair of photochemical lesions. Two main types of repair processes are recognized: photoreactivation and dark repair (see reviews by Setlow, 1966 and 1967, and Witkin, 1966 and 1968).

Photoreactivation requires intense white light irradiation *after* ultraviolet irradiation, blue wavelengths being most effective. The photoreversal of mutations in cellular systems has been reported many times. In a study of forward mutants at the *ad-3* locus of *Neurospora,* Kilbey and de Serres (1967) observed that both complementing (base pair substitution) and non-complementing (possibly frameshift) mutations were equally photoreversible, in general agreement with Drake's (1966d) results. In a subsequent study, however, Kilbey (1967) discovered one mutation out of 15 whose ultraviolet-induced reversion was not photoreversible.

The interpretation of photoreversal data is frequently complicated by indirect effects (Weatherwax, 1961; Kaplan, 1963; Witkin, Sicurella, and Bennett, 1963; Jagger, Wise, and Stafford, 1964; Witkin, 1964; Kaplan and Witt, 1965). If *E. coli* is irradiated with white light *before* being irradiated with ultraviolet light, the yield of both lethal and mutational lesions is reduced; this process is called photoprotection. In *phr* strains of *E. coli*, which are unable to carry out ordinary photoreactivation (Harm and Hillebrandt, 1962), limited photoreactivation (probably identical to photoprotection) of both lethal and mutational lesions is still possible, so long as another cellular repair system (dark repair, to be described below) is still functioning. The action spectra of direct and indirect photoreactivation are different (Kondo and Jagger, 1966; Kondo and Kato, 1966). Indirect photoreactivation therefore appears to promote the action of some other repair system, probably by introducing damages which delay DNA replication, thus permitting a longer interval for dark repair systems to operate. Witkin (1964) observed indirect photoreactivation of several auxotrophic mutations, but not of mutations resistant to streptomycin; the latter were also insusceptible to dark repair.

The yield of mutants following ultraviolet irradiation depends strongly upon the post-irradiation treatment of the cells. If irradiated cells are incubated in saline before being plated, both lethal and mutational lesions tend to disappear (liquid holding recover); some kind of metabolic activity is required in order to prevent "mutation frequency decline" (Doudney and Haas, 1958, 1959, 1960;

Phillips, 1961; Clarke, 1967). Mutation frequency decline is now usually interpreted as resulting from repair occurring before DNA replication permanently establishes mutations (Witkin, 1964; Harm, 1966b). A definite role of repair processes in the modification of ultraviolet mutagenesis in *E. coli* was established by Hill (1965), who observed that mutation and lethality were strongly and proportionately enhanced in an ultraviolet-sensitive mutant. (Mutants of this type have been given various names, such as *uvr, hcr,* and B*s*. Many different genes are involved in maintaining the fully resistant state in *E. coli*.) Dark repair and mutation frequency decline are both inhibited by caffeine and acridines (Witkin, 1961; Lieb, 1961; Doudney, White, and Bruce, 1964; Shankel and Kleinberg, 1967; Clarke, 1967), although the two drugs may act by separate mechanisms. An additional repair system insensitive to caffeine is claimed to have been revealed in *hcr* mutants, but its existence is in dispute (Kneser, Metzger, and Sauerbier, 1965; Kneser, 1966; Harm, 1966a, 1967).

A number of reports have appeared which suggest that two different classes of ultraviolet-induced mutations can be distinguished in bacteria. Kaplan and Kaplan (1956) reported that dry cells were much more efficiently mutagenized in an environment of low relative humidity, and that such mutations were not photoreactivated; mutations induced at a high relative humidity were readily photoreactivated. Witkin and Theil (1960) and Witkin (1964) observed mutation frequency decline and indirect photoreactivation following the induction of revertants of various auxotrophs, but not following the induction of streptomycin-resistant mutations. Clarke (1967) detected different effects of caffeine on lethality and mutation, and suggested that one type of lesion leads both to mutation and to lethality, while another leads only to mutation. Zelle, Ogg, and Hollaender (1958) observed that mutations at the *Sm* and *pur* loci were induced with single-hit and multiple-hit kinetics, respectively. Bridges, Dennis, and Munson (1967) also observed that suppressors and true revertants of an ochre mutation were induced with different kinetics, and were differentially affected by repair; mutation decline was observed only for suppressors. However, both types were susceptible to photoreactivation. Finally, moderately high frequencies of deletions have been observed to result from ultraviolet mutagenesis in bacteria, in contrast to bacteriophage T4 (Cook and Lederberg, 1963; Demerec, Gillespie, and Mizobuchi, 1963).

Certain characteristics of ultraviolet mutagenesis in bacteria have been recorded, whose relevance to the underlying mechanisms is somewhat obscure, but will eventually require explanation. Belser (1961) observed that the mutation frequency was sharply increased in magnesium-depleted cells. Jensen and Haas (1963) irradiated *B. subtilis,* extracted the DNA, and assayed mutations in it by transformation; the mutations were recoverable from the irradiated cells only after about one DNA doubling time. Howarth (1965, 1966) reported that the *col I* episome conferred both an increased resistance to lethality, and an increased sensitivity to mutagenesis in *Salmonella.* Finally, a number of workers have observed that induced mutations often appear to arise in pure mutant clones, as if both strands of DNA were mutated simultaneously. This problem will be discussed in Chapter 16.

The ultraviolet photochemistry of the nucleic acids has been reviewed by Smith (1966) and by Setlow (1966, 1967). Attention has centered on the pyrimidines, since they are chemically modified from ten to one hundred times more frequently than are the purines. It is still perfectly possible for the purines to be important targets, however, because they may be altered in biologically important ways which have yet to be discovered, and because they may transfer absorbed energy to pyrimidines. The photochemistry of the pyrimidines depends very strongly upon whether they are irradiated free in solution or incorporated into polynucleotides. Two major types of product are obtained: hydrates and dimers (Fig. 14-1).

The 4,5 double bond of pyrimidines can be hydrated to form 4-hydroxy-5-hydro derivatives. Thymine hydrate probably reverses very rapidly under most conditions, and has not been recovered from irradiated samples. The hydrates of uracil and cytosine are stable under normal biological conditions, though the cytosine hydrate may be slowly reversed by heat or acid. A fraction of the reversed cytosine hydrate appears in the form of uracil (Johns, Le Blanc, and Freeman, 1965), particularly at high pH. Cytosine hydrates form rapidly in denatured DNA, but less rapidly in native DNA (Grossman, 1968). The hydrate may nevertheless be important in mutagenesis since mutations are infrequent events, compared to lethal lesions, and only a few unsuppressed cytosine hydrates might be required; it may also be important because both replicating DNA and bacteriophage DNA within the virus particle probably contain denatured regions. In addition to pyrimidine hydrates, 4,5-dihydrothymine has recently been detected in irradiated DNA (Yamane, Wyluda, and Shulman, 1967). Furthermore, when uracil is irradiated in the presence of HCN, 5-cyanouracil is formed (Evans, Hidalgo-Salvatierra, and McLaren, 1967).

A variety of pyrimidine dimers are produced when polynucleotides are irradiated, those containing thymine being most common. Of the three types of dimers thus far detected (Figure 14-1), the cyclobutane type (Beukers and Berends, 1960b) has been most intensively studied; the TOT (Pearson, Ottensmeyer, and Johns, 1965) and TNC (Wang and Varghese, 1967) species have not been fully characterized as yet. [Two more possible thymine dimer configurations are described by Varghese and Wang (1968) and by Stafford and Donnellan (1968).] A considerable number of isomers of cyclobutane dimers are possible, but those which form in polynucleotides appear to arise from a pair of adjacent pyrimidines in the same polynucleotide chain. However, a few of the dimers may form between residues nearly opposite each other on the complementary chains, producing crosslinks. Cyclobutane dimers are both formed and reversed by ultraviolet irradiation; equilibrium is strongly on the dimer side at about 280 nm, and strongly on the monomer side at 240 nm. Cytosine residues in cyclobutane dimers tend to undergo deamination to form uracil. When cells are irradiated, the possibility may also exist for dimer

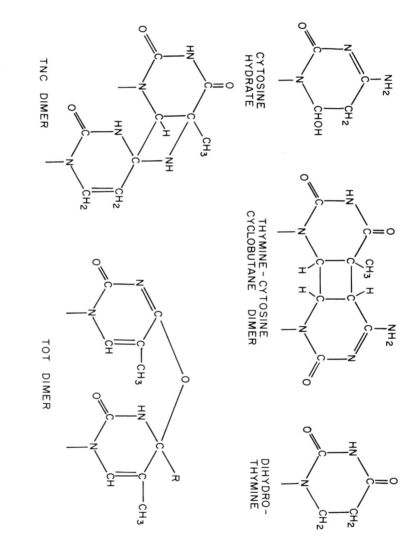

FIGURE 14-1. *Some photoproducts detected in ultraviolet-irradiated nucleic acids.*

formation between incorporated and unincorporated pyrimidine residues, in much the same way that psoralen dimerizes with pyrimidines (see following section on photodynamic mutagenesis, and Fig. 14-3).

The mechanism of formation of cyclobutane dimers depends strongly upon the conditions of irradiation. Dimerization of simple solutions of the nucleosides proceeds via an excited triplet state, and is sharply inhibited by paramagnetic species such as oxygen and manganese (Beukers and Berends, 1960a; Rahn, Shulman, and Longworth, 1965; Greenstock et al., 1967). Dimerization in polynucleotides probably proceeds via an excited singlet state (Eisinger and Shulman, 1967; Greenstock et al., 1967). Methods have been developed recently which offer the possibility of inducing thymine dimer formation in DNA without the formation of any other photochemical product (Lamola and Yamane, 1967; Greenstock and Johns, 1968).

Ultraviolet-induced excitations in DNA may migrate up and down the molecule (Weill and Calvin, 1963; Pearson et al., 1966), and if proflavin molecules are intercalated in the DNA, they may act as energy traps. As a result, the formation of pyrimidine dimers is sharply depressed by proflavin, leading to decreased mutagenesis and enhanced levels of survival (Buekers, 1965; Webb and Petrusek, 1966; Setlow and Carrier, 1967; Setlow and Setlow, 1967). In organisms able to incorporate azathymine into DNA, thymine dimer formation and lethality are also suppressed (Wacker, 1963; Günther and Prusoff, 1962). However, these modifying factors have not yet been exploited in studies of mutational mechanisms.

Pyrimidines have also been shown to form dimers with cysteine (such as 4-hydro-5-S-cysteine-uracil), and may thereby become involved in crosslinks between nucleic acids and proteins (Smith, 1964; Smith and Aplin, 1966; Smith, Hodgkins, and O'Leary, 1966).

In addition to the moderately well defined photoproducts discussed thus far, a number of additional products have been surmised to occur on the basis of chromatographic spots (see Smith, 1966, for references), action spectra (Cavilla and Johns, 1964), and effects of low pH and anaerobiosis (Sauerbier and Haug, 1964; Haug and Sauerbier, 1965).

Effects of ultraviolet irradiation have been studied in two types of *in vitro* polymerizing systems. When polyuridylic acid is used as a template for polyphenylalanine synthesis, irradiation inactivates its priming ability and induces serine incorporation (Grossman, 1962, 1963; Wacker, Jacherts, and Jacherts, 1962). Irradiation at 285 nm produces only inactivation, while irradiation at 245 nm produces both inactivation and serine incorporation. Heating the irradiated polyuridylic acid reduces serine incorporation. These results strongly suggest that uracil dimers inactivate the template, while uracil hydrates induce a *phe → ser* coding change which corresponds to a $U \to C$ transition. The uracil hydrate presumably assumes the enolic form (see Figure 10-1).

When polycytidylic acid is used as a template for polyguanylic acid synthesis, irradiation inactivates its priming ability. The polymerization of GTP can be partly restored by the addition of ATP, which is also incorporated (Ono, Wilson, and Grossman, 1965). Heating again reverses these effects, indicating the formation of cytosine hydrates which can mimic uracil, possibly because they assume the imino form (Figure 10-1). (The authors suggest that an ionized species may be responsible for the mispairing.)

The two main classes of repair systems—photoreactivation and dark reactivation—turn out to possess rather different enzymatic mechanisms. Photoreactivation splits cyclobutane dimers *in situ* (Setlow, Boling, and Bollum, 1965), and does not release bases or break the DNA backbone. The photoreactivating enzyme, which has been isolated and assayed *in vitro* (Rupert, Goodgal, and Herriott, 1958; Wulff and Rupert, 1962), binds to irradiated DNA and is released by white light irradiation. It is not yet clear whether cyclobutane dimers are the only substrates for the enzyme, or whether it may more generally unsaturate 4,5 single bonds in pyrimidines. The TNC type of dimer, for instance, appears to be reversed by photoreactivation (Wang and Varghese, 1967), and there are some indirect indications that even pyrimidine hydrates may be reversed (Tao, Small, and Gordon, 1967). Ultraviolet-induced DNA crosslinks are also photoreactivable (Marmur and Grossman, 1961).

The dark repair systems do not split cyclobutane dimers, but instead excise them from DNA, leaving an exposed single-stranded region (Setlow and Carrier, 1964, 1966; Boyce and Howard-Flanders, 1964). DNA repair generally ensues, with the result that a burst of synthesis is observed after ultraviolet irradiation of competent cells (Pettijohn and Hanawalt, 1964). A number of genetic loci must cooperate to achieve excision and repair (Howard-Flanders, Boyce, and Theriot, 1966; Howard-Flanders and Boyce, 1966). It is not yet clear whether the systems which excise cyclobutane dimers also excise other types of altered bases.

A surprising interaction between excision repair and mutagenesis has recently been revealed. First, many thymine cyclobutane dimers may persist without lethality in replicating DNA (Rupp and Howard-Flanders, 1968; Sauerbier and Hirsch-Kauffmann, 1968). They may, however, induce the appearance of persistent gaps in progeny molecules, and in addition, may cause the appearance of mutations in later generations (Bridges and Munson, 1968; but see Witkin, 1969). These gaps may eventually be repaired by either of two processes (Witkin, 1968, 1969): a rapid and relatively inaccurate synthesis (in which case mutations are produced in high frequencies), and a slow but very accurate synthesis. The rapid and mutagenic type of repair appears to be under the control of genes determining UV sensitivity: certain UV-sensitive strains of *E. coli* and of *Schizosaccharomyces pombe* are vastly less sensitive to UV mutagenesis than is the wild type (Witkin, 1968; Miura and Tomizawa, 1968; Nasim, 1968). The nature of the "fast but sloppy" system is not yet resolved.

The diverse information available concerning the mutational and photo-chemical actions of UV irradiation does not easily lend itself to interpretation in terms of simple mechanisms. The $C \rightarrow T$ transitions which predominate in viruses, for instance, are attractively interpreted as the results of hydration of cytosine, leading to tautomerization and the formation of a thymine analogue. However, two types of evidence make this simple interpretation improbable: 5HMC (in bacteriophage T4) is probably much less susceptible to hydration than is ordinary cytosine, and in addition, when mutations are induced in T4 using an energy transferring photosensitizer which only affects thymine, $G:C \rightarrow A:T$ transitions are produced in an undiminished yield (Marvin Meistrich, personal communication). Both base pair substitutions and frameshift mutations may therefore arise because of the vagaries of certain types of repair synthesis, as is suggested by the results obtained in *E. coli*. This hypothesis also helps to explain the photoreversal of both types of mutations.

PHOTODYNAMIC ACTION

A variety of dyes can sensitize organisms to white light irradiation, with the result that both lethal and mutational hits are produced (Table 14-2 and Figure 14-2). Mutation in the bacterial systems usually follows complex kinetics; although sometimes interpreted as multiple-hit mutational processes, the kinetics may merely indicate progressive changes in the accessibility of the target molecules as a function of time of irradiation. Böhme and Wacker (1963), for instance, observed that mutations were produced with very different kinetics in two closely related strains of *P. mirabilis,* and Mathews (1963) observed drastic effects upon the cell membrane and cell permeability.

Nakai and Saeki (1964) report that oxygen was required with their dyes, mutagenesis being abolished by flushing with nitrogen. Oxygen is also usually required for photodynamic inactivation (Welsh and Adams, 1954), except in the case of psoralen (Mathews, 1963).

In a number of instances, mutations have been induced in bacteria by white light irradiation in the absence of added dyes (Kelner and Halle, 1960). Kubitschek (1967) observed that the rate of this process was decreased about two-fold under anaerobic conditions, and suggested that two separate processes were involved. Webb and Malina (1967) observed that wavelengths above 380 nm required oxygen to be effective, whereas the near-ultraviolet wavelengths (355 nm) were quite effective in the absence of oxygen. Webb, Malina, and Benson (1967) determined the action spectrum of the white light process, and concluded that riboflavin could be the chromophore. For the time being, at least, it is therefore reasonable to suppose that white light mutagenesis in the absence of added dyes is not fundamentally different from that produced in the presence of added dyes.

Ritchie (1964, 1965) observed that most mutants photodynamically induced in bacteriophage T4 were induced to revert by base analogues, and concluded that they contained transitions. Rather small numbers of mutants were studied, however, and some were reverted neither by base analogues nor by proflavin. A larger collection of T4 mutants was studied by Drake and McGuire (1967b), who

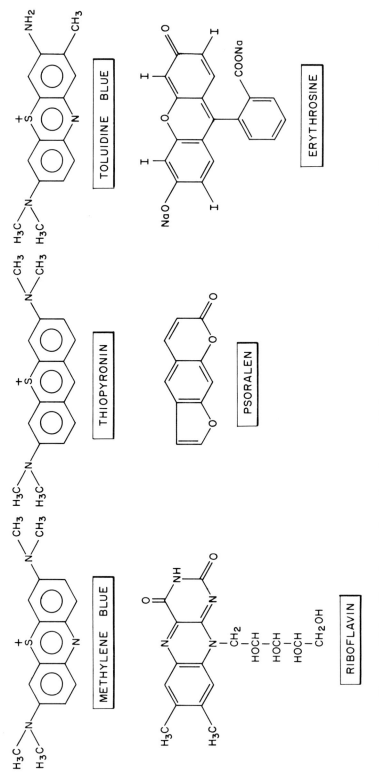

FIGURE 14-2. *Dyes frequently used as photosensitizing agents. For the structure of proflavin, see Figure 12-5.*

TABLE 14-2

PHOTODYNAMIC MUTAGENESIS

Organism	Mutation	Dye	Reference
Serratia marcescens	aberrant colonies	ER	Kaplan (1949, 1956)
Penicillium notatum	aberrant colonies	ER	Kaplan (1950b)
Sarcina lutea	penicillin resistance	MP,TB	Mathews (1963)
Proteus mirabilis	$phe \rightarrow phe^+$	MP,TP	Böhme and Wacker (1963)
Escherichia coli	resistance to T7	ER	Kaplan (1950a)
	resistance to T5	AO	Webb and Kubitschek (1963)
	$try \rightarrow try^+$	AO,AY,MB,TB	Nakai and Saeki (1964)
Bacteriophage kappa	plaque morphology	MB	Brendel and Kaplan (1967)
Bacteriophage T4	$r^+ \rightarrow r$	PF	Ritchie (1964, 1965)
		PS,TP	Drake and McGuire (1967b)
		(many)	Calberg-Bacq, Delmelle, and Duchesne (1968)
	$rII \rightarrow r^+$	MB	Barricelli and Del Zoppo (1968)
		MB	Brendel (1968)

AO = acridine orange; AY = acridine yellow; ER = erythrosin; MB = methylene blue; MP = 8-methoxypsoralen; PF = proflavin; PS = psoralen; TB = toluidine blue; TP = thiopyronin.

observed that proflavin-revertible frameshift mutations were produced weakly or not at all, and that about a third of the mutants which were produced could not be reverted by base analogues. These presumably contain transversions; and as a result, many of the mutants which could be induced to revert by base analogues also probably contain transversions. About half of the mutants which responded to base analogues also responded to hydroxylamine, suggesting that they contained G:C base pairs at the site of the mutational lesion.

A variety of types of T4rII mutants, including frameshifts and both types of transitions, have been shown to be photodynamically reverted (Barricelli and Del Zoppo, 1968; Brendel, 1968). These results, together with the results of Drake and McGuire (1967b), indicate that photodynamic mutagenesis produces a variety of types of microlesions. The kinetics of the process need further investigation, however, since Calberg-Bacq, Delmelle, and Duchesne (1968) observed that T4r forward mutations were produced in a single-hit manner at a rate of about 7×10^{-4} mutants per lethal hit (in agreement with the value of about 4×10^{-4} reported by Drake and McGuire), whereas Brendel (1968) observed that T4rII mutants were reverted in an approximately two-hit manner.

The main photodynamic reaction promoted by most of these dyes is the oxidative degradation of guanine (Simon and Van Vunakis, 1962; Wacker, Turck, and Gestenberger, 1963; Sussenbach and Berends, 1963; Simon and Van Vunakis, 1964; Bellin and Grossman, 1965; Sussenbach and Berends, 1965; Dellweg and Opree, 1966). Thymine also reacts, although to a much lesser extent (Simon and Van Vunakis, 1962). The degradation of guanine results in a variety of products (Wacker, Turck, and Gestenberger, 1963; Sussenbach and Berends, 1965), one of which is parabanic acid (Figure 11-1), and another of which might be a thymine analogue (by extrapolation from the results of Matsuura and Saito, 1967). The degraded guanine residues are generally unable to function in messenger RNA (Simon, Grossman, and Van Vunakis, 1965), and may eventually lead to strand scission (Freifelder, Davison, and Geiduschek, 1961; Freifelder and Uretz, 1966). In addition to degradation, however, guanine can experience coupling with solutes in its vicinity; the presence of tris buffer, for instance, leads to the formation of at least two tris adducts (Van Vunakis et al., 1966). The number of possible photodynamic products may therefore be immense.

Psoralen and its derivatives behave very differently from the other dyes described in Table 14-2 and Figure 14-2 (Wacker et al., 1964; Krauch, Drämer, and Wacker, 1967; Musajo, Bordin, and Bevilacqua, 1967; Musajo et al., 1967). It reacts mainly with pyrimidines by the formation of cyclobutane addition products such as those illustrated in Figure 14-3, and shows no oxygen requirement. The reaction with cytosine promotes deamination to form the uracil addition product.

The mechanism of photodynamic mutagenesis remains obscure, partly because of our rather rudimentary understanding of the relevant nucleic acid photochemistry, and partly because of indications that the known photo-

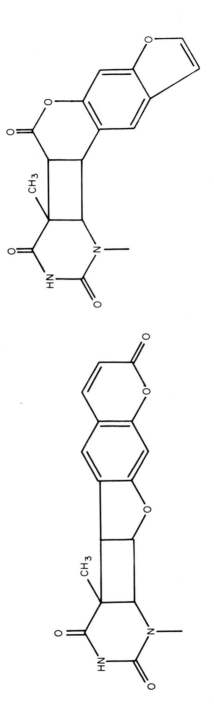

FIGURE 14-3.   *Psoralen-thymine dimers induced photodynamically.*

chemical reactions may not apply. As is often the case with mutagens, many lethal hits occur for each mutational hit, and the main photochemical reactions may be relevant only to the former. Thus Drake and McGuire (1967*b*) observed that about 25 lethal hits occur in T4 for each mutational hit (extrapolating from the rate of induction of *r* mutations), and if repair processes operate upon any of the primary photochemical lesions, then the ratio of lesions to mutations may be even greater. Frequent repair of photodynamic lesions is in fact indicated by the results of Singer and Fraenkel-Conrat (1966), Brendel and Kaplan (1967), and Böhme and Geissler (1968). Furthermore, although psoralen and thiopyronin appear to promite radically different types of photochemical reactions, their mutational specificities are indistinguishable (Drake and McGuire, 1967*b*).

A transversional mispairing scheme involving parabanic acid was described in Chapter 11, but this does not explain the formation of transversions induced by psoralen. Transitions might be produced by reactions which promote either the enolization or the ionization of guanine or of thymine. If the psoralen-cytosine complex could act as a base during DNA replication, then deamination of cytosine to uracil would also produce transitions. It seems very likely, however, that most of the mutationally important photodynamic reactions have yet to be described.

HEAT

Very high temperatures applied to bacteria or spores produce high frequencies of mutations (Zamenhof and Greer, 1958; Zamenhof, 1960). When *E. coli* cells were heated at $60°$, or were dried and heated at $135°-155°$, they were inactivated roughly five decades, while the survivors contained as many as 10 percent mutants. The conditions which are effective for mutagenesis also induce depurination (Greer and Zamenhof, 1962), but for the reasons discussed in previous chapters, it is doubtful that depurination alone is sufficient to induce mutations.

High temperatures at neutral pH have not to my knowledge been reported as mutagenic for viruses. However, a spontaneous mutational process in T4 has a high Q_{10} and would proceed very rapidly at high temperatures (Drake, 1966*b*; Drake and McGuire, 1967*a*); heating at $44°$ seems to produce mutations by the same process (Krieg, personal communication). This type of mutation usually arises at sites which are also mutated by hydroxylamine, but it appears to induce both G:C → C:G transversions and G:C → A:T transitions (and perhaps other base pair substitutions as well). Krieg also observed that the mutational specificity of heat at neutral pH did not correlate with the mutational specificity of acid, even though both treatments induce depurination.

15. Spontaneous mutation

Spontaneous mutants undoubtedly arise from a multiplicity of causes, and contain a multiplicity of different mutational lesions. The types of mutations and their characteristic frequencies depend upon the organism, upon the genes surveyed, and to a certain extent upon environmental and other modifying factors. In addition, different mechanisms of spontaneous mutation may predominate depending upon whether the organism is dormant or dividing.

A semantic problem always hovers in the background when mechanisms of spontaneous mutation are under discussion. Suppose, for example, that a major fraction of the spontaneous mutations which arise in an organism are found to be caused by some component of the environment such as background ionizing radiation. If the organism were maintained under conditions of extremely low background irradiation, for instance in certain locations deep underground, its spontaneous mutation rate would drop. Should the mutations appearing in the normal habitat still be considered spontaneous? Consider also the question of the mistakes attributable to errors in DNA synthesis: if two species of DNA polymerase exist, one being characterized by a lower error rate, are we to assume that the less accurate polymerase is mutagenic? For purposes of simplicity and consistency, it is best to take the mutation rate characteristic of an organism in its natural habitat, or under a set of standard laboratory conditions, to be its spontaneous mutation rate. Fortunately, spontaneous mutation rates of *E. coli* and bacteriophage T4 are relatively insensitive to variations in the growth medium, ionic environment, and other similar variables, so long as inhibitory extremes are avoided.

REPLICATION-DEPENDENT MUTATION

The molecular configuration of mutations which arise in bacteriophage T4 have been more closely surveyed than in any other organism, and it is therefore appropriate to emphasize this virus. T4 is not considered to be fully representative, however, for it is already clear that the relative frequencies of various types of mutations differ in other viral and bacterial systems. The great majority of mutations to be found in a new T4 stock have arisen by replication-dependent processes. The yield of mutations is insensitive to growth temperatures over 26° to 42°, and to the composition of the medium: a simple synthetic medium such as M9 and a complete medium such as broth supplemented with yeast extract (and various intermediate combinations) produce virtually identical mutational yields.

When T4*r* mutations are collected, they divide among the *r*I, *r*II and *r*III categories (see Chapter 7) as indicated in Table 15-1. The *r*III mutations are infrequent, and have not been characterized as to type of molecular lesion. The *r*II mutations have been characterized by mapping experiments and reversion analysis, with the results illustrated in Table 15-2.

TABLE 15-1

APPROXIMATE RELATIVE FREQUENCIES OF SPONTANEOUS T4*r* MUTANTS*

Genotypic Class	Replication- Dependent	Replication- Independent
*r*I	0.35	0.43
*r*II	0.64	0.55
*r*III	0.01	0.02

* The *r*II class contains leaky mutants which plate on K(λ) cells but map within the *r*II region; some authors have grouped these with the phenotypically similar *r*III mutants. Sources: Benzer, 1957; Drake, 1966c and d; Drake and McGuire, 1967a.

TABLE 15-2

APPROXIMATE RELATIVE FREQUENCIES OF SPONTANEOUS T4*r*II MUTANTS*

Class	Replication- Dependent	Replication- Independent
Deletion	0.06	0.00
Frameshift	0.72	0.12*
BA(+)	0.19	0.59
Other	0.03	0.29

* Deletion mutants are detected by their lack of reversion and by mapping experiments. Frameshift mutants are detected by their susceptibility to reversion by acridines; the frequency listed in the second column(*) is probably an over-estimate of a virtually null class. Mutants described as BA(+) are susceptible to reversion by base analogues, and contain transitions and sometimes transversions. "Other" mutants are not reverted by acridines or base analogues, and are probably transversions. Sources: Benzer, 1957, 1959; Drake, 1966c, d; Drake and McGuire, 1967a; Freese, 1959a; Orgel and Brenner, 1961; Tessman, 1962.

The two most prominent T4*r*II hot spots which appear in Figure 5-5, the spontaneous mutational spectrum, contain frameshift lesions. If these hot spots were reduced in magnitude or abolished, a situation which is already approximated in T6 (Benzer, 1961a), then the ratio of frameshift mutations to base pair substitutions would be reduced from 3.3 to 1.6. If, in addition, the frequency of detection of base pair substitution mutations in the *r*II region were increased to 50 percent from its actual level of below 4 percent (for example, by increasing the functional activity required of the *r*II cistrons; see Chapter 5), then the ratio of frameshift mutations to base pair substitutions would be further reduced to 0.12. The types of mutational lesions which are produced and detected, therefore, depend both upon gene composition and upon circumstances unrelated to intrinsic mutation rates.

The importance of DNA polymerase accuracy in organism-specific spontaneous mutation frequencies is vividly illustrated by two examples. First, an amber mutation in bacteriophage λ has been observed to revert one hundred times more rapidly when the virus is replicating lytically (under its own direction) than when it is replicating in the lysogenic state as an element of the host chromosome (Dove, 1968). Second, the mutation rate throughout the T4 genome can be either increased or decreased by mutations in the T4 DNA polymerase gene; this subject will be discussed more fully below.

Striking differences in mutational behavior between different cistrons have also been observed in the T4*r* system. When *r* mutations of the base pair substitution type are induced by agents such as base analogues, the relative frequency of the *r*I class rises from the background level of about one-third, to as high as 70 percent.

When induced by proflavin, however, relatively few *r*I mutations appear compared to *r*II mutations. These differences might result from a markedly greater efficiency of detection of base pair substitution mutations in the *r*I cistron compared with the *r*II cistrons.

The kinds and frequencies of mutations which arise at different points within a cistron presumably depend upon the local base pair composition, but the factors which determine mutation rates remain obscure. Hot spotting, the most vivid type of local variation in mutability, is generally more pronounced for spontaneous than for induced mutation (Benzer, 1961*a*; Krieg, 1963*a*). This is strikingly illustrated in the case of proflavin, which induces exclusively frameshift mutations: the two largest T4*r*II spontaneous hot spots, which contain frameshift mutations, are virtually unrepresented in the proflavin spectrum. An experimental attack upon the nature and extent of neighboring base effects on local mutation rates is long overdue. A beginning has, however, been made toward an *in vitro* analysis of neighboring base pair effects upon base selection, at least in RNA synthesis (Goldberg and Rabinowitz, 1961; Slapikoff and Berg, 1967). Adenine in the template directs the incorporation of both uridine and its analogue pseudouridine, for instance, but if both pyrimidines are available at the same time, uridine tends to be incorporated at positions where the preceding base is another pyrimidine, whereas pseudouridine tends to be chosen if the preceding base is guanine. In addition, Pitha, Huang and Ts'o (1968) have demonstrated base selection effects upon errors in poly-U-directed stacking of adenine.

Mutation rates at different points within the cistron may depend both upon local base pair composition, and upon natural chemical modifications of the

DNA molecule which are now known to occur. The λ chromosome, for instance, shows large variations in the (G+C)/(A+T) ratio along its length, and easily denatured (presumably A:T-rich) regions can be observed (Inman, 1967); it will be of considerable interest to measure mutation rates within and outside of such regions. The T4 chromosome is characterized by specific patterns of glycosylation of the 5HMC residues (Burton et al., 1963), and patterns of methylation of the bases (Gold et al., 1964; Hausmann and Gold, 1966; Sellin, Srinivasan, and Borek, 1966); the influence of these substitutions upon mutation rates is currently under examination.

Numerous attempts have been made to detect differences in mutation rates in various cistrons in the repressed and derepressed states, since RNA transcription might sensitize the DNA to damage. A difference has been observed only once: both the spontaneous and the EMS-induced reversion rate of mutations in the *E. coli* glycerol kinase gene are reported to be several fold higher under derepressed (constitutive) than under repressed conditions (Koch, 1968).

REPLICATION-INDEPENDENT MUTATION

The question of whether the nonreplicating gene is subject to spontaneous mutation is an old one. Mutations have been shown to accumulate in stored gametes of higher forms (see review by Muller, Carlson, and Schalet, 1961), in dry spores of *Neurospora* (Auerbach, 1959), in dry but not wet spores of *Bacillus subtilis* (Zamenhof, Eichhorn, and Rosenbaum-Oliver, 1968), and in starved or stationary phase *E. coli* (Ryan, 1956; Ryan, Nakada, and Schneider, 1961). The results in gametes and *E. coli*, but probably not in *Neurospora* or *Bacillus subtilis*, were complicated by cryptic DNA synthesis, which may take the form of slow DNA replication dependent upon the breakdown of other molecules, or of DNA repair synthesis. Grigg and Stuckey (1966) observed that stationary phase mutation in *E. coli* is suppressed by caffeine, strongly suggesting that repair synthesis promotes such mutation.

An unequivocal answer to the question of whether nonreplicating genes can mutate was finally provided by studies with bacteriophage T4 (Drake, 1966*b*; Drake and McGuire, 1967*a*); neither repair nor replication occurs in free virus particles. Certain *r*II mutants were observed to accumulate revertants linearly with time during long intervals of storage. Only mutants susceptible to reversion by hydroxylamine were affected, indicating that the G:C base pair is the mutational target for the process. The rate was strongly temperature-sensitive, being about 20-fold greater at 20° than at 0°. The forward mutation rate ($r^+ \to r$) was about 3×10^{-7} per particle per day at 0°, and about 3×10^{-6} per particle per day at 20°; translated into total mutations per virus, these values become approximately 0.01 and 0.1 per particle per year, respectively. While low compared to the spontaneous mutation rate during intracellular replication, these rates nevertheless become highly significant when one considers the problem of genome storage over years or decades. Methods of reducing these rates, other than by reducing the temperature, have not been explored.

The properties of a collection of *r* mutants leisurely accumulated over 1.3 years at 20° are summarized in Tables 15-1 and 15-2. Many of the mutants are revertible neither by base analogues nor by proflavin, and therefore appear to contain transversions; as a result, many of the mutants revertible by base analogues also probably contain transversions. Nearly all of the mutants which were induced to revert by base analogues were also induced to revert with hydroxylamine. Since G:C base pairs were originally established as the mutational targets, the major mutational pathway appears to be G:C → C:G. This conclusion is supported by the observation that amber (UAG) mutants, but not ochre (UAA) mutants, revert when held at 44° for one week (D. R. Krieg, personal communication.

The mechanism of this process is unknown, but there are indications that different mechanisms may operate in different organisms. In *E. coli*, for instance, repair synthesis seems to be mandatory. Furthermore, Auerbach (1959) observed that *Neurospora* spores stored at 4° and then shifted to 30° accumulated mutations much faster than did spores stored continuously at 30°; no such dual mechanism was observed in T4.

MUTATOR GENES

From time to time, organisms are encountered which exhibit spontaneous mutation rates differing from the values characteristic of the species. The altered genetic stability can itself often be mapped as a genetic character. In this section, alterations of mutational stability which encompass the entire genome will be considered; these are commonly attributed to mutator genes when they involve increased mutation rates. Genetic instabilities confined to a single gene or a single site will be considered in the next section.

The numerous instances of proven or surmised mutator genes in higher organisms will not be discussed, since there is little hope of understanding their mechanisms of action at the present time. However, important advances in the understanding of mutator genes have recently been achieved in microorganisms.

The most profitably investigated bacterial mutator gene was discovered in *E. coli* by Treffers, Spinelli, and Belser (1954), and mapped near the *leu* region by Skaar (1956). Studies by Bacon and Treffers (1961) suggested that the *mutT* mutator acts with considerable specificity, since it promotes only one of a number of pathways of reverse mutation of an auxotroph. The details of this specificity were determined by analysis of mutations induced by the mutator in the tryptophan synthetase A protein (Yanofsky, Cox, and Horn, 1966; Cox and Yanofsky, 1967). Specifically, only those amino acid substitutions compatible with A:T → C:G transversions were observed. This is a remarkable result in two ways. First, it is the only example of a purely transversional mechanism of mutation. Second, it is the only example of a mutational process operating exclusively upon the A:T base pair. The effect of the Treffers mutator gene on reversion rates is prodigious, frequently producing increases on the order of

10^3-fold. The action of this unusual mutational pressure over 1200-1600 bacterial generations is claimed to reduce the A:T base pair frequency by about 0.3 percent, and to reduce the frequency of thymidylate runs. The antimutagenic effects of spermine and quinacrine extend to mutations produced by *mutT* (Johnson and Bach, 1966). An essentially identical mutator has recently been selectively isolated by Helling (1968).

The mechanism of action of *mutT* is not known, but it is tempting to suppose that it results from an alteration in the accuracy of the DNA polymerase. Bacteriophage T3, which probably produces its own DNA polymerase, is not affected by *mutT*, whereas bacteriophage T7, which may not produce its own DNA polymerase, is affected (see Hausmann and Gomez, 1967). Since bacteriophage T4 is fully dependent upon its own DNA polymerase, it should not be affected by *mutT*. Cox (personal communication) did not detect an effect of *mutT* upon the reversion rate of a variety of T4rII mutants, nor upon the forward mutation rate, but Pierce (1966) reported that the Treffers mutator did in fact mutagenize T4. However, T4 mutation rates were increased only about 1 percent as strongly as were *E. coli* mutation rates, suggesting that *mutT* is acting circuitously in this case, for instance, by producing abnormal cells or metabolites which in turn produce mutations.

Two other *E. coli* mutators have been at least partially characterized. The *ast* mutator (Goldstein and Smoot, 1955) produces several percent auxotrophs in a single colony. It maps near *proA*, possibly near *mutT* (Zamenhof, 1966), and also affects virulent bacteriophages reproducing in an *ast* host cell (Zamenhof, 1967). The *mutS1* mutator (Siegel and Bryson, 1964, 1967) maps near *argB* and has an additive effect with *mutT*. It produces base pair substitutions, but is indifferent to an excess of exogenous normal bases, suggesting that an enzyme of DNA replication is involved, rather than an abnormal base. A number of additional *E. coli* mutators have been isolated, but have been studied in less detail (Jyssum, 1960; Mohn and Kaplan, 1967; Kada, 1968).

A mutator gene has been described in the LT7 strain of *Salmonella typhimurium* (Demerec et al., 1957; Miyake, 1960). It maps near *purA*, and has been claimed to produce an abnormal base (Kirchner and Rudden, 1966). Base pair substitutions, mostly transitions, are produced (Kirchner, 1960), and recombination frequencies are sometimes affected as well (Balbinder, 1962). Selective devices have recently been employed to isolate strong mutators in *Neisseria meningitidis* (Jyssum and Jyssum, 1968) and in *B. subtilis,* where the isolate was also temperature sensitive for DNA synthesis (Gross, Karamata, and Hempstead, 1968).

Mutator activity has frequently been observed to result from mutations which increase UV sensitivity or decrease recombination (Böhme, 1967; Jyssum, 1968; Mohn, 1968; Prozorov and Barabanshchikov, 1967 and 1968; von Borstel, 1968; von Borstel et al., 1968; Zakharov, Kozhina, and Fedorova, 1968). The yeast mutator described by von Borstel is especially interesting, since it appears to produce frameshift mutations. As the biochemistry of these mutants becomes clear, they may provide important insights into mutational mechanisms.

A purely viral mutator system was discovered by Speyer (1965) in bacteriophage T4, where mutator activity results from certain temperature-sensitive lesions which map in gene 43, the structural gene for T4 DNA polymerase (de Waard, Paul, and Lehman, 1965; Warner and Barnes, 1966). This observation provided the first experimental results in support of the notion that DNA polymerase participates directly in the process of base selection during DNA replication (see Koch and Miller, 1965; and Kornberg, 1969). The two mutants most closely studied by Speyer, *ts*L56 and *ts*L88, produce a variety of types of base pair substitution mutations, including both transitions and transversions (Speyer, Karam, and Lenny, 1966; Speyer et al., 1967); frameshift mutations are not produced at unusual rates. At least 5 of 40 gene 43 *ts* mutations, and certain *amber* mutations as well, result in mutator activity. In extensive tests of about fifty gene 43 *ts* mutants, Drake et al. (1969) estimated that at least 80 percent produced mutator activity.

Both the *ts*L56 and the *ts*L88 DNA polymerases have been studied *in vitro* (Hall and Lehman, 1968), but neither was observed to be dramatically more error-prone than the wild type polymerase. The *ts*L56 enzyme produced $G \rightarrow T$ transversions about 4-fold more frequently, and various other substitutions no more frequently, than the wild type enzyme. Since these error frequencies are far smaller than those observed in genetic studies, it seems impossible at present to reconstruct *in vitro* the behavior of T4 DNA polymerase.

In a genetic study of *ts*L56, Freese and Freese (1967) observed a complete absence of interaction between mutator-induced mutagenesis and chemical mutagenesis, using base analogues, hydroxylamine, EMS, and nitrous acid: the probability of mutation was the sum of the probability due to mutator activity plus the probability due to chemical mutagenesis. The *ts*L56 mutator activity, therefore, does not appear to affect the process of base recognition by the DNA polymerase. Speyer (1969), however, observed a moderate synergism between base analogue and *ts*L56 mutagenesis, which has been confirmed by Albrecht and Drake (unpublished).

A number of mutations in gene 43 exhibit a remarkable *antimutator* activity (Drake and Allen, 1968; Drake et al., 1969). The two sites most intensively studied, *ts*CB87 and *ts*CB120, strongly and specifically reduce the frequency of transitions, but do not affect the frequency of transversions or of frameshift mutations. The antimutators suppress some but not all types of chemical mutagenesis, in striking contrast to the behavior of the mutator *ts*L56.

The behavior of T4 gene 43 lesions suggests an insight into the mechanism of replication-dependent spontaneous mutation. If DNA polymerase does in fact take an active role in base selection, then it must contain binding sites for both parental-strand and progeny-strand bases. Without specifying in any detail the enzymatic mechanism of DNA polymerization or the mechanism of information transfer, it is nevertheless possible to distinguish two different ways in which errors could arise. The first hypothetical error mechanism depends upon incorrect binding, for instance of an adenine residue in a site which should bind guanine. A mutation which changed the binding site error rate would probably produce a change in the frequency of transitions but not necessarily in the frequency of transversions, since it is unlikely that a purine would occupy a pyrimidine binding site or vice versa. The altered mutation rate would also probably extend to certain types of chemical mutagenesis, since the probability of accepting a chemically altered base would be changed. The T4 antimutator lesions behave as expected of lesions which change the binding-error rate. The second hypothetical error mechanism depends, not upon incorrect binding, but upon perturbed communication between binding sites. An erroneously constructed DNA polymerase might result, for instance, in the association of two adenine sites, with the consequent production of A:T → T:A transversions. Mutations which change the accuracy of communication between binding sites might change the rates characteristic of a number of different mutational pathways, and both transitions and transversions could result. Since the accuracy of the binding sites would not necessarily be affected, this type of mutant DNA polymerase could be indifferent to chemical mutagensis. The tsL56 mutator behaves as expected of such a lesion, and the $E.$ $coli$ $mutT$ mutator probably does also, since it produces transversions. It is clearly a task for the future to determine in detail the properties of additional mutants of the DNA polymerase.

(A model of base selection very different from the above is proposed by Kornberg [1969]. It assumes that base selection is made first by base pairing, and that the polymerase then accepts or rejects the inserted residue.)

LOCALIZED GENETIC INSTABILITIES

Specific genes or sites sometimes suddenly become highly mutable, without a concomitantly increased mutability throughout the genome. A number of plausible molecular explanations for such instabilities may be imagined, but none have yet been established in any particular instance. Mutations which revert at extremely high rates could result from duplications which are reversed by recombinational events. Mutations could also occur which would not by themselves inactivate a gene, but which would nevertheless greatly increase its chance of mutation to an inactive form. Base pair substitutions, for instance, might assist in the creation of new hot spots, in which case most of the mutations which appeared later would be localized at a single site. Base pair substitutions could also partially inactivate a gene, not sufficiently to produce a phenotypic change in the organism under standard conditions, but sufficiently to render visible many further mutations within the cistron which might otherwise have gone undetected. In this case an entire gene might be unstabilized.

Detailed analyses of localized instabilities have been initiated in a few microbial systems. Nasim (1967) and Loprieno et al. (1968) described instabilities in the yeast *Schizosaccharomyces pombe* which were localized within a single member of a pathway of five genes. Riyesaty and Dawson (1967) observed the unstabilization of a cistron in *Salmonella typhimurium* which resulted in the appearance of mutations at many sites within the cistron. A number of instances of localized instability in bacteria have been shown to be associated with an episomal element. This is particularly clear in the case of instabilities which may be "cured" by acridines (Gundersen, Jyssum, and Lie, 1962; Gunderson, 1963, 1965; Schwartz, 1965; Sompolinsky et al., 1967). In certain cases sites on the bacterial chromosome seem to have been the targets of the unstabilizing process, and in other cases episomal genes themselves were affected. A number of instances of highly unstable revertants originally suspected of arising under episomal control now seem likely to have arisen from a chromosomal but extracistronic suppressor which was intrinsically unstable, or else was strongly selected against during cell growth (Hill, 1963; Dawson and Smith-Keary, 1963).

16. Mutational heterozygotes

Newly arisen mutant individuals frequently appear in the form of heterozygotes (often called mosaics in cellular organisms). Heterozygosity becomes evident as a result of the subsequent segregation of two different types of homozygous progeny, which in microorganisms often leads to distinctive plaque or colony morphologies. Most proposed mutational mechanisms predict the formation of heterozygotes. However, some newly arisen mutants appear to be homozygotes, and this observation has induced imaginative speculations concerning the mechanism of DNA duplication.

BASE ANALOGUE HETEROZYGOTES

The heterozygotes which appear as intermediates in 5-bromouracil mutagenesis (Pratt and Stent, 1959) have already been mentioned briefly in Chapter 10. When bacteria were infected with T4rII transition mutants and then exposed to the analogue, the earliest of the induced revertants to mature were usually r^+/r heterozygotes. Revertants matured at later times were usually homozygous, presumably as a result of DNA replication. Mutants induced in the forward direction also arise as heterozygotes. The obvious and straightforward interpretation of these results assumes the existence of a G:BU intermediate which then segregates G:C and A:BU (= A:T) progeny.

Strigini (1965) studied the induction of mutations by 5-bromouracil during the course of a cross between a pair of outside markers. He observed that the mutants were unusually frequently heterozygous for the outside markers, whereas heterozygosity was not expected to extend beyond the immediate site of the mutated base pair. Two explanations for this result are possible. One is that the mutations arose during repair synthesis accompanying recombination (see Figure 3-4), and that the outside markers were included within recombinational heterozygotes; this hypothesis also requires that repair synthesis be somewhat less accurate than normal DNA replication. Another explanation, suggested by the results of Abe and Tomizawa (1967), is that 5-bromouracil stimulates the initiation of new replication forks, resulting in the clustering of mutations at the ends of molecules. These mutations would therefore experience an elevated probability of inclusion within a region of terminal redundancy. Although the mutational lesion itself would be a heteroduplex heterozygote, the outside markers would frequently appear in the form of terminal redundancy heterozygotes. The lengths of many of the heterozygotes involving the outside markers were much longer than is characteristic of recombinational heterozygotes, but were easily compatible with terminal redundancy heterozygotes; thus the second explanation is favored.

Witkin and Sicurella (1964) studied the production of heterozygotes in *E. coli*, with results strikingly different from those obtained using bacteriophage T4. The cells were grown for one generation in bromouracil, and then plated on a medium which rendered visible the *lac* mutant and nonmutant portions of the resulting colonies. Segregation in *E. coli* is invariably more complex than in T4, since in addition to the duplex nature of the DNA molecule, many of the cells are multinucleate. In the present case, half of the cells contained two or more nuclear bodies, and half of the induced mutants arose in mosaic colonies. Thus multinucleate cells were sufficiently frequent to account for all of the observed heterozygosity, and the additional mosaicism expected from the segregation of G:BU heteroduplex heterozygotes was completely absent. Furthermore, nuclear lethality was insufficient to account for the lack of mosaicism. Explanations of this result will be considered below.

MUTATIONAL HETEROZYGOTES
IN FREE VIRUS PARTICLES

The *in vitro* induction of mutational heterozygotes was strikingly demonstrated by I. Tessman (1959) in a study which compared nitrous acid mutagenesis in bacteriophages T4 and ΦX174. In ΦX174, which contains single-stranded DNA, all of the mutants arose as homozygotes, whereas in T4, most of the mutants arose as heterozygotes. In addition to being consistent with the strandedness of the viral DNA, the mutational data also indicate that the chemically altered bases produced by nitrous acid mispair during subsequent DNA replication with a probability close to unity.

Green and Krieg (1961) studied the action of the alkylating agent EMS upon bacteriophage T4 (see Chapter 13). In this case the mutants appeared with a clonal distribution which indicated that they arose with an approximately equal probability at each replication of the mutated strands. In contrast to nitrous acid, the chemically altered bases produced by alkylation appear to mispair with a rather low probability during subsequent DNA replications. EMS produces mosaics in bacteria as well as in viruses (Loveless, 1958; Loveless and Howarth, 1959).

Most mutagens acting upon free T4 particles produce mutational heterozygotes; these include nitrous acid and hydroxylamine (Vielmetter and Wieder, 1959; Schuster and Vielmetter, 1961; Freese, Bautz-Freese, and Bautz, 1961), ultraviolet irradiation (Drake, 1966c), and photodynamic mutagenesis (Drake and McGuire, 1967*b*). Spontaneous mutants arising in the absence of DNA replication are also heterozygous (Drake and McGuire, 1967*a*). In *E. coli*, however, quite different results have been obtained: mutations induced photodynamically and also by base analogues usually appear as homozygotes (Witkin, 1964; Kubitschek, 1964, 1966).

Mutational heterozygosity has interesting consequences when it occurs in viral genes which must act before the onset of viral DNA replication. If chemical mutagens usually mutate only one strand of DNA at a given site, and if the base sequence of messenger RNA is copied from only one of the DNA strands, then the messenger RNA encoded by mutational heterozygotes will be wild type only when copied from a nonmutant strand of DNA. For instance, if wild type T4 particles are mutagenized *in vitro* and are then used to infect bacteria which do not permit the growth of *r*II mutants, then particles mutated in the *r*II region of the transcribed DNA strand will abort, whereas particles mutated in the complementary strand will multiply and produce both mutant and wild type progeny. If the treated particles infect permissive cells, however, all of the induced mutants will be recovered. Thus Tessman (1962*a*) observed that the yield of T4*r*II mutants induced by nitrous acid and recovered following growth in nonpermissive cells, was only half as great as that obtained following growth in permissive cells. Levisohn (1967) obtained complementary results when he studied the reversion of *r*II mutants by hydroxylamine. About half of the mutants yielded a maximum frequency of revertants when plated directly on nonpermissive cells, whereas the remainder only yielded a maximum frequency of revertants when first grown in permissive cells and then plated on nonpermissive cells. The first class of mutants presumably contained 5HMC residues on the transcribed strand, whereas the second class presumably contained guanine residues.

ACRIDINE HETEROZYGOTES

T4*r* mutations induced by proflavin first appear as heterozygotes (Drake, 1964). Strigini (1965) studied the reversion of frameshift mutations during the course of a cross between a pair of outside markers, each about 10 map units away from the site of reversion. Many of the new revertants turned out to be heterozygous for one or both of the outside markers. The length of these heterozygotes requries either that they be extraordinarily long heteroduplex heterozygotes, or that they be terminal redundancy heterozygotes. It is possible to distinguish these two types by studies of forward mutations. T4*r*II heteroduplex heterozygotes should be able to multiply in nonpermissive cells with a probability of 0.5, if the mutations are randomly distributed between the transcribed and the complementary DNA strands; whereas terminal redundancy heterozygotes should always be able to multiply in nonpermissive cells, since an intact *r*II region will be present at the nonmutated end of the chromosome (Hertel, 1965). Recent experiments by Lindstrom (unpublished) confirm these expectations, and also demonstrate that proflavin-induced heterozygotes are able to multiply in nonpermissive cells with a probability of 1.

According to the Streisinger hypothesis concerning the origin of frameshift mutations (see Chapter 12), the most probable configuration for these heterozygotes is that of a heteroduplex heterozygote located within a region of terminal redundancy (and hence of terminal heterozygosity). The feasibility of joint heteroduplex terminal redundancy heterozygotes has already been demonstrated by Hertel (1963), who showed that a single T4 particle may sometimes carry three different alleles of the same site: wild type, deletion mutation, and point mutation.

MASTERSTRAND HYPOTHESES

Although mutational heterozygosis is the rule in bacteriophage T4, it is clearly the exception in some bacterial systems. Several schemes of DNA replication have therefore been proposed, all of which have a common property: the sequences of *both* DNA daughter strands are determined by the sequence of only *one* of the DNA parental strands—the masterstrand. As a result, a mutation which occurs in the masterstrand produces exclusively mutant progeny chromosomes, while a mutation which occurs in the complementary strand is not transmitted to any of the progeny chromosomes, although it might affect the phenotype of the parent. Although masterstrand schemes stipulate a skewed flow of genetic information, they must be consistent with the segregation pattern of parental and progeny atoms which is characteristic of semiconservative DNA replication.

The first such scheme (Jehle et al., 1963; Jehle, 1965) was created to explain, within the framework of a mechanically satisfactory model, the high accuracy of DNA replication compared to the low accuracy of hydrogen bonding specificity; its masterstrandedness was incidental. In its most simple form, this model stipulates that one parental strand specifies the base sequence of the complementary daughter strand, which in turn specifies the base sequence of its own complement (Figure 16-1). In another scheme, Kubitschek and Henderson (1966) proposed that the pair of bases to be added to the pair of growing points at the replication fork arrive simultaneously in a hydrogen-bonded configuration. The selection of the correct pair of bases in this model depends upon the DNA polymerase, which in turn takes instructions from the two parental bases. Although both parental strands normally cooperate in specifying the progeny residues, a mismatched pair of parental bases (such as A:C) obviously cannot do so. Instead, only the strand with $3' \rightarrow 5'$ bonding operates in the selection process, and thereby becomes a masterstrand (Figure 16-1). A special masterstrand mechanism has also been proposed for viruses (Barricelli and Womack, 1965; Barricelli, 1965). This model assumes that one strand of the injected DNA is inactivated, but that DNA replication later proceeds in the usual semiconservative manner.

SCHEMATIC SUMMARIES OF MASTERSTRAND MODELS

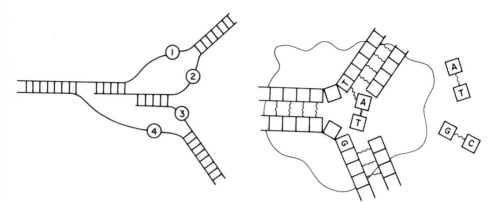

FIGURE 16-1. *Hypothetical mechanisms for the determination of both daughter strands by a single parental strand in DNA replication. Left: the Jehle model, in which strand 1 directs the synthesis of strand 2, which in turn directs the synthesis of strand 3, which ultimately pairs with strand 4. Right: the Kubitschek-Henderson model, in which a heteroduplex heterozygote (the guanine-thymine base pair) directs the synthesis of one homozygous and one heterozygous progeny molecule.*

Before considering the evidence bearing on masterstrand hypotheses, it should be noted that by no means have all the possible models yet managed to find their way into print. A mechanism might be imagined, for instance, in which a random decision was made at each site about which base would be the masterbase. Furthermore, it is not necessary to specify a particular mechanism of DNA replication in order to consider the problem of information transfer during DNA replication; in fact, the reverse approach promises to be more fruitful in the long run.

Werner's (1968) analysis of T4 DNA replication suggests that the distribution of growing points early in infection is assymmetric, with the result that a heteroduplex heterozygote would initially segregate many more copies of one homozygote than of the other. As a result, heterozygotes might sometimes not be recognized. Since mutational heterozygotes are very common in T4, however, this effect is not directly relevant to the masterstrand problem.

The outstanding alternatives to masterstrand schemes as explanations of mutational homozygosity are repair and lethality. Since repair of mismatched bases might easily be unable to discern which base is mutant and which is wild type, heterozygotes might be converted with equal frequency into homozygous mutant and homozygous wild type. Lethality can be imagined at two levels: nuclear lethality and single-strand lethality. A number of studies now strongly suggest that repair and lethality are, in fact, adequate explanations of mutational homozygosity.

The rather special case of strand inactivation proposed by Barricelli conflicts with the common observation that most T4 mutants arise in mixed clones. A second assumption is therefore required: during subsequent replications, chemically altered bases sometimes mispair, but, in addition, sometimes pair normally (Barricelli, 1965; Rosner and Barricelli, 1967). While this condition is clearly satisfied in alkylation mutagenesis, it is very probably not true for nitrous acid mutagenesis: exclusively homozygous mutant clones are induced in the single-stranded bacteriophage ΦX174. Furthermore, the T4 mutant clone size distribution is at least partly explained as the result of randomizing factors in the replication and maturation of chromosomes within the vegetative pool.

Some of the mutations induced in T4 *in vitro* do appear as homozygotes, the fraction increasing slowly with increasing doses. Freese and Freese (1966) have presented evidence which strongly suggests that these homozygotes are the result of repair. They introduced mutations into T4 using hydroxylamine, an agent which produces relatively few lethal hits. They then introduced lethal hits using ultraviolet irradiation, an agent which produces relatively few mutations. A considerable fraction of the lethal hits introduced by ultraviolet irradiation are repaired by an excision system (see Chapter 14), and the excision usually includes many more bases than those directly involved in the photochemical lesion. As a result, the frequency of mutations recovered from hydroxylamine mutagenesis was observed to decline with increasing doses of ultraviolet irradiation, and at the same time, the relative frequency of mutations recovered as heterozygotes decreased even more rapidly. When a T4 mutant which was incapable of excising photochemical lesions was used, the conversion of mutational heterozygotes ceased completely. It is therefore very likely that the few mutational homozygotes which are observed normally are due to repair, although the specific systems involved have not yet been established.

Kubitschek (1966) tested two strains of *E. coli* deficient in the excision repair of ultraviolet lesions, but found that 2-aminopurine and photodynamic mutagenesis still produced mutational homozygotes in them. Since repair of mutational lesions in *E. coli* may occur by systems entirely different from those involved in the repair of photochemical damage, these results are insufficient to distinguish between masterstrand and repair mechanisms. However, a probable example of repair of heterozygosity is observed in *E. coli* infected with bacteriophage λ, which, in contrast to bacteriophage T4, is generally susceptible to host repair systems. Kellenberger, Zichichi, and Epstein (1962) observed that c^+/c heteroduplex heterozygotes are usually very difficult to detect in λ. If the particles infect irradiated host cells, however, the heterozygotes are readily observed, suggesting that repair systems which would normally operate on the virus particles are now tied up by photochemical lesions in host DNA.

Similar results have been obtained in certain DNA transformation systems. Doerfler and Hogness (1968) separated the complementary strands of bacteriophage λ DNA by density gradient centrifugation, making use of the small natural density difference between the two strands. They then constructed *am/am*[+] heterozygotes by annealing the appropriate pairs of strands. The two reciprocal heterozygotes (am^+ on the transcribed or on the complementary strand) were then compared in a DNA transformation system. When nonpermissive cells were used, both types of hybrid were able to multiply with a probability of 0.5, even though the amber mutation resided in a gene whose expression was required before DNA replication. When irradiated cells were used, only one of the two reciprocal heterozygotes was able to multiply, suggesting that repair of viral lesions was inhibited by irradiation of the host cell. However, the significance of these results is lessened by the observation that the method failed entirely in a second gene. Gurney and Fox (1968) studied the uptake of density-labelled transforming DNA into pneumococci., Only single strands were integrated into the recipient chromosome, and these formed density hybrids. After one generation, no markers from the donor DNA could be detected in DNA of pure recipient (light) density, whereas conversion to homozygosity would have resulted in the appearance of donor markers in this fraction. Guerrini and Fox (1968 a, b) observed that virtually all the transformants were exclusively heterozygous, and were converted to homozygotes by a single replication. When repair was stimulated by treating newly transformed cells with mitomycin, however, evidence of conversion of heterozygosity to homozygosity was obtained. In pneumococci, therefore, some heteroduplex heterozygotes appear to survive quite efficiently; others, however, may usually be rapidly repaired (Louarn and Sicard, 1968). In B. $subtilis$, different markers show very different frequencies of conversion to homozygosity, and marker pairs tend to be jointly repaired if sufficiently close together (Bresler, Kreneva, and Kuschev, 1968).

Nasim and Auerbach (1967) observed that mutagenic treatments of the yeast $Schizosaccharomyces$ $pombe$ which resulted in high levels of survival produced high frequencies of mosaics among induced mutations. The relative frequency of mosaics decreased with increasing doses of mutagen. The quantitative aspects of the results were interpreted to be inconsistent both with masterstrand mechanisms and with the production of nonmosaic colonies by lethal sectoring; the pure mutant colonies which were observed were believed to result from repair or from mutagenic lesions involving both DNA strands. Haefner (1967a, b) and Guglieminetti (1968) approached the problem of strand lethality directly by micromanipulative separation of the descendents of mutagenized cells in the first several generations. They concluded that strand lethality frequently occurred, accounting for most or all of the pure-mutant colonies observed both by them and by Nasim and Auerbach. However, both Haefner and Auerbach (1967) emphasized that a residue of pure-mutant clones which cannot be explained by strand lethality occur. Since a masterstrand mechanism is rendered unlikely by the frequent occurrence of mosaic colonies, repair processes are likely to be involved. This supposition is supported by the observation that in certain UV-sensitive strains, both the mutation rate and the frequency of mosaicism are increased (Nasim, 1968).

A critical test of the masterstrand hypothesis has recently been performed using bacteriophage λ (Russo, Stahl, and Stahl, personal communication). When the cross *sus*O X *c sus*P is performed, the *sus*$^+$ recombinants can be selected by plating on cells nonpermissive for amber mutants; among these recombinants, *c*/*c*$^+$ recombinational heterozygotes (see Chapter 3) are detected by their mottled plaque morphology. The lysate from the cross was passaged in cells and medium which were density-labelled using $C^{13}N^{15}$; conditions were chosen which slowed λ DNA replication. The progeny particles were then banded in a density gradient (see Figure 3-3). They formed three bands: light ("free riders" which were rematured before being replicated), hybrid, and heavy (totally progeny strands). The light band contained the same frequency (about 25 percent) of *c*/*c*$^+$ heterozygotes among *sus*$^+$ recombinants as did the original cross lysate. Comparison of the relative sizes of the hybrid and heavy bands indicated that an average of three DNA generations had elapsed. A viable masterstrand model for this system would have to assume that either DNA strand (different strands at different replications) could be the masterstrand in order for segregation (mottled plaques) to occur in the first place. The reader may then verify for himself that after *n* generations, the heterozygote frequency will be reduced by 2 in both the hybrid and heavy bands. However, not $25/2^3 = 3$ percent, but fewer than 0.3 percent of heterozygotes appeared in these bands. Thus, the replication of heterozygotes appears to produce only homozygotes, and a central prediction of the masterstrand hypothesis fails to be realized.

17. Suppression

When a mutant reverts because of the appearance of a second mutation at a new site, the second-site mutation is called a suppressor. Thus the active combinations of (+) and (−) frameshift mutations discussed in Chapter 12 constitute pairs of mutual suppressors. Suppressor mutations may occur either inside or outside of the originally mutated cistron, and suppressors of viral mutations may even occur in the host genome. Certain mutations may also be overcome by drugs of exogenous origin, a process of chemical suppression called phenotypic reversion. The various types of suppressors and their mechanisms of action have been most recently reviewed by Gorini and Beckwith (1966).

Genetically, suppressor mutations are conveniently categorized according to whether or not they fall within the originally mutated cistron. Intracistronic suppressors occur (by definition) at sites distinct from the originally mutated site, but may fall within the same codon. (When reversion occurs at the original site, but still produces a nonwild genotype, then a distinct term such as pseudoreversion must be used; see line 7 of Figure 17-1.) Extracistronic suppressors can be somewhat arbitrarily classified according to their mechanisms of action (see Gorini and Beckwith, 1966). We shall limit the discussion of extracistronic suppressors to those which induce misreading of messenger RNA, thus producing "phenotypic mutations," but we will not consider those suppressors which provide a substitute for the required function [for example, Emrich (1968), Itikawa et al. (1968), and Kuo and Stocker (1969)].

Whatever their location, suppressor mutations are usually detected by backcrosses to the wild type which result in the reappearance of the original mutant. Although suppression often produces a phenotype which is distinct from the true wild type, many instances of suppression have been recorded in which an apparently wild phenotype was in fact observed.

INTRACISTRONIC SUPPRESSION

The special properties of intracistronic suppression involving frameshift mutations have been considered in detail in Chapter 12, and need not be recounted here. It should be emphasized, however, that as a general rule, intracistronic

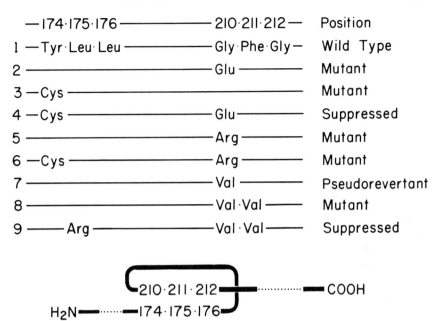

FIGURE 17-1.   *Amino acid replacements in the E. coli tryptophan synthetase A protein. The residue involved is indicated in the top line, and the phenotypes of the nine different versions of the enzyme are given at the right. "Mutant" enzymes are completely inactive, whereas pseudorevertants and suppressed mutants usually produce enzymes with intermediate levels of activity. The hypothetical interaction between amino acid residues 174-175-176 and 210-211-212 is sketched at the bottom.*

suppression of a frameshift mutant will only be brought about by another frameshift mutation, and intracistronic suppression of a base pair substitution mutant will only be brought about by another base pair substitution. Rare exceptions to this rule may occur when the mutational lesions affect signals for chain initiation or chain termination (Sarabhai and Brenner, 1967; Riyasaty and Atkins, 1968).

Although high frequencies of false revertants (putative instances of suppression due to the appearance of a semiwild phenotype) have been recorded in many systems, backcrosses with sufficient resolving power to detect intracistronic suppressors have not often been performed. Jinks (1961) observed that each of 129 h^+ revertants of various T4*h* mutants resulted from suppression, and all of the suppressors which were mapped fell within the original mutated gene. Drake (1963*b*, 1964) observed the frequent appearance of false revertants among both spontaneous and induced revertants of T4*r*II mutants; the suppressors which were mapped all fell within the *r*II region, some very close to the first site and others much farther away. These results indicate that intracistronic suppressors can easily outnumber true revertants, depending upon the particular gene or sites studied.

Intracistronic suppression has been profitably studied in the *E. coli* tryptophan synthetase A protein, where a large fraction of revertants are clearly distinct from the original wild type (Allen and Yanofsky, 1963). One quite complex pattern of suppression is illustrated in Figure 17-1 (Helinski and Yanofsky, 1963; Yanofsky, Horn, and Thorp, 1964). The amino acid substitutions shown in polypeptides 2, 3, and 4 of the figure demonstrate intracistronic suppression between two sites (numbers 174 and 210) 36 amino acid residues apart. When a mutation occurred at position 212, it too could be suppressed by a mutation 36 residues away, at position 176 (polypeptides 8 and 9). It is therefore likely that the polypeptide assumes a tertiary configuration somewhat as suggested in the bottom portion of the figure, with the mutually suppressing amino acid residues in apposition.

Intracistronic suppression may also occur by a mechanism which produces a single amino acid difference between the wild type and the suppressed revertant: intracodon suppression. This process is also well documented by mutations observed in the tryptophan synthetase A protein (see Figure 10-4). The pathway from wild type to mutant to suppressed revertant may, for instance, occur by the amino acid pathway from glycine to arginine to threonine. The corresponding codon changes are GGA → AGA → ACA: a mutation in the first position has been suppressed by a mutation in the second position. The reversion of nonsense codons frequently proceeds by the substitution of an amino acid different from the wild type, sometimes by intracodon suppression and sometimes by pseudoreversion.

MUTATIONAL MODIFICATIONS OF TRANSLATION

A mutationally altered polypeptide may regain its function by a suppressor mutation which brings about a second-site amino acid substitution, or by an extracistronic suppressor mutation which still brings about an amino acid substitution at the originally mutated site. The second type of suppression often results from processes which induce mistakes in translation, resulting in amino acid insertions in the polypeptide which are not encoded in the corresponding messenger RNA. If the translational mistake level is sufficient to restore a critical level of gene function, but does not at the same time inactivate too many of the other proteins in the cell, the cell will survive. Several different components of the translation apparatus might be altered to promote such mistakes: the amino acid activating enzymes, the transfer RNA molecules, and the ribosomes. Since these components are usually coded by cellular genomes, the suppression of viral mutations by translational modifications not only represents extracistronic suppression, but also extragenomic suppression. It should be noted, however, that large viruses such as T4 may encode occasional components of the translational apparatus (Neidhardt and Earhart, 1967; Weiss et al., 1968).

The suppression of the nonsense codons, UAG (amber), UAA (ochre), and UGA, must occur either by intracodon or by extracistronic suppression. A variety of extracistronic suppressors have been discovered. These are character-ized phenotypically by two parameters: the amino acid which they insert into the chain-terminating codon, and the frequency of insertion (Table 17-1). The suppressors map at widely separated loci in *E. coli* (Signer, Beckwith, and Brenner, 1965). At least some of these suppressors represent altered transfer RNA species, since chain termination measured during the *in vitro* translation of RNA viruses or messenger RNA from bacteriophage T4 can be overcome by adding transfer RNA from cells carrying suppressor mutations (Engelhardt et al., 1965; Capecchi and Gussin, 1965; Gesteland, Salser, and Bolle, 1967; Söll, 1968). The complete base sequence of both a wild type and an amber-suppressor tyrosine transfer RNA have recently been determined (Goodman et al., 1968). The transfer RNA anticodon (Figure 17-2), which is complementary to the messenger RNA codon and runs in the opposite direction, was altered by a $G \rightarrow C$ transversion in the suppressor strain, resulting in a $GUA \rightarrow CUA$ anticodon change. The new anticodon had lost the ability to recognize the tyrosine codon, but had gained the ability to recognize the amber codon. The tyrosine codon, however, can still be translated by another transfer RNA encoded by *E. coli*.

TABLE 17-1

Suppressor	Suppressed Codons			Amino Acid Inserted	Probability of Chain Propagation
	UAG	UAA	UGA		
I	+	−	−	serine	0.6
II	+	−	−	glutamine	0.3
III	+	−	−	tyrosine	0.5
IV	+	−	−	lysine	0.1
B	+	+	−	?	0.05
C	+	+	−	?	0.01
4	+	+	−	?	0.01
UGA	−	−	+	?	0.6

Sources: Stretton, Kaplan, and Brenner, 1966; Kaplan, 1967; Sambrook, Fan, and Brenner, 1967.

Although the tyrosine suppressor clearly arose by a base pair substitution, nonsense suppressors probably also result from frameshift mutations. A class of "supersuppressors" in yeasts very strongly resembles bacterial nonsense suppress-ors (see, for instance, Gilmore, Stewart, and Sherman, 1968). Supersuppressors arise specifically during meiosis, and are not inducible by agents which produce base pair substitutions (Magni, von Borstel, and Steinberg, 1966; Magni and Puglisi, 1966). As indicated in Chapter 12, these characteristics strongly suggest that the mutations result from base pair additions or deletions. Nonsense suppressors have also been induced in *Salmonella* by the frameshift-specific mutagen ICR-191 (Fink, Klopotowski, and Ames, 1967). It is therefore likely

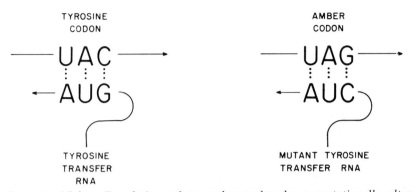

FIGURE 17-2.   *Translation of an amber codon by a mutationally altered tyrosine transfer RNA.*

that anticodons can be changed by the addition of a base, or even by the deletion of a base which forces an adjoining base to join the anticodon.

Suppressors of amber and UGA codons can achieve a high probability of amino acid insertion at the mutant site. Ochre suppressors, on the other hand, are characteristically weak, and reside in slowly growing cells which are easily overgrown by nonsuppressing revertants. Since chain-terminating codons probably punctuate the end of translation of most cistrons, an efficient suppressor will link together many cellular proteins translated from polycistronic messenger RNA molecules. It is therefore likely that the UAG and UGA codons are used only infrequently for chain termination in *E. coli.*

Missense suppression has also been profitably studied in the tryptophan synthetase A protein system. The A36 mutation (glycine to arginine) is suppressed by a mutation which results in the production of two species of A protein, a majority of mutant molecules containing arginine and about 5 percent of suppressed molecules containing glycine (Carbon, Berg, and Yanofsky, 1966); the same suppressor permits the growth of a set of bacteriophage T4 mutants (Reid and Berg, 1968). In an *in vitro* system, suppression depends upon the presence of transfer RNA from suppressor cells. Since high levels of exogenous arginine do not affect the efficiency of suppression, it was suggested that a glycine transfer RNA had been modified to occasionally recognize the arginine codon, rather than that an arginine transfer RNA had been modified to occasionally accept glycine. The A78 mutation (glycine to cysteine) is suppressed by a mutation which leads to the insertion of glycine in about 2 percent of the A protein molecules (Gupta and Khorana, 1966). The *in vitro* analysis again indicated that transfer RNA is responsible for the suppression. Since high levels of exogenous cysteine reduce the efficiency of suppression, it seems likely that a cysteine transfer RNA molecule has been altered so that it is occasionally loaded with glycine. Whereas the A36 transfer RNA may well have been mutated within the anticodon, the A78 transfer RNA is likely to have been mutated in the region recognized by the amino acid activating enzyme.

Additional missense suppressors which substitute aspartic acid and valine residues (probably by the insertion of glycine and glutamine, respectively) have been described in the A protein system (Brody and Yanofsky, 1963; Berger and Yanofsky, 1967). Bacterial mutations which suppress *ts* mutations in λ have also been described (Regös and Szende, 1967); *ts* mutants are very likely to contain missense mutations, but the particular amino acids are unknown.

While the suppressors described above appear to produce altered transfer RNA molecules, other suppressors produce altered ribosomal components. A strain of *E. coli* carrying a *ts* lesion produces temperature-sensitive 50S ribosomal subunits and a concomitant missense suppressor activity (Apirion, 1966). The corresponding amino acid substitution is not known. Another strain of *E. coli* carries a mutation called *ram*, which affects the 30S ribosomal subunits and which leads to suppression of all three chain terminating codons (Rosset and Gorini, 1969).

Selection for revertants or for mutants with altered patterns of resistance to streptomycin has often resulted in the simultaneous recovery of auxotrophic mutations at other, completely unrelated loci. Although no mechanisms have yet been established, some of these systems probably result from alterations in transcription or translation which result not only in a profitable misreading of the mutant codon, but also in a potentially disastrous misreading of a normal codon in another gene. Mutations at streptomycin sensitivity loci (which probably affect ribosomal structure) have produced or revealed auxotrophy at a variety of loci in *Proteus mirabilis* (Böhme, 1961b), in *Salmonella typhimurium* (Watanabe, 1960; Goldschmidt, Matney, and Bausum, 1962), and in *Bacillus subtilis* (Kohiyama and Ikeda, 1960). Revertants for one type of auxotrophy have produced or revealed auxotrophy for other growth factors in *Serratia,* and reappearance of the primary auxotrophy occurred with the simultaneous disappearance of the secondary auxotrophies (Kaplan, 1961). Similar examples of joint mutation have been observed in *Neurospora* by Grigg and Sergeant (1961). In *Salmonella*, reversion of *leu* mutants frequently occurs by distant suppressor mutations, some of which induce secondary *cys* and *try* auxotrophs (Mukai and Margolin, 1963). In this system, however, translational components are not involved. Instead, suppression results from simple mutational inactivation of some cistron, and when this occurs by deletion, nearby cistrons (such as *cys* and *try*) may also be deleted.

SUPPRESSORS IN REVERSION ANALYSIS

Much of the information which has been obtained concerning the configuration of mutational sites and the specificity of chemical mutagens, has been derived from studies of induced reversion. Although frameshift mutants are usually suppressed intracistronically by other frameshift mutations, and base pair substitution mutants by other base pair substitutions, more specific deductions about mutated sites often cannot be made unless the revertant can be shown to be identical to the wild type. In a number of viral and bacterial systems, ever more exhaustive tests to detect false revertants have revealed ever increasing proportions of them among collections of revertants. As a result, rigorous criteria are required for deciding when a revertant has regained the exact wild genotype.

A number of criteria are both theoretically promising and technically feasible. These include both backcrosses and phenotypic tests, eventually encompassing determinations of amino acid sequence. Backcrosses with resolving power sufficient to detect intracodon recombination by means of selective methods for the recovery of mutant phenotypes are usually possible in viral and bacterial systems, but are limited to a certain extent by the possibility of anomalously low recombination frequencies between extremely close markers (see Chapter 4). Phenotypic tests may take many forms, depending upon the physiological characteristics of the system under study, but visual examination of plaque or colony morphology, and tests for temperature sensitivity in the restored function, have frequently revealed a high proportion of all false revertants detected. Allosteric interactions such as feedback inhibition are often much more sensitive to mutational alterations than are catalytic activities of enzymes, and are therefore sensitive indicators of false reversion. Except for the complications arising from codon degeneracy, however, the most sensitive test available for the detection of intracistronic suppression is to determine the primary structure of the protein. However, at present this is usually a very difficult task.

Measurements of forward mutation rates at the original site have also been used as sensitive indicators of the revertant DNA configuration (Benzer, 1961a). This method has the potential advantage of sensitivity to the exact base pair composition of a region, even when codon degeneracy might complicate the interpretation of amino acid analyses. As a general method, however, it is clearly very difficult.

It is not technically feasible, and perhaps not even possible in principle, to demonstrate rigorously which members of a set of revertants have regained the exact wild genotype. Fortunately, however, estimates of mutational specificity are usually based upon analyses of the properties of many mutants at different sites. Although some of the mutants in such studies will probably be misclassified, it is likely that the majority will be correctly classified, and that the corresponding mutational pathways will therefore be revealed.

PHENOTYPIC REVERSION

A number of agents can restore the functional activity of mutated genes without directly substituting the required function itself. Certain of these agents affect the accuracy of translation, thereby bringing about nonheritable phenotypic mutations. The first such agent to be studied in detail was 5-fluorouracil. This compound is readily incorporated into messenger RNA, and causes the production of defective proteins, including temperature-sensitive alkaline phosphatase (Naono and Gros, 1960) and enzymatically inactive β-galactosidase (Bussard et al., 1960). (A very similar result was obtained by Hamers and Hamers-Casterman [1961] using thiouracil.) It is likely (by analogy with 5-bromouracil mutagenesis of DNA) that 5-fluorouracil acts by producing nonheritable U → C transitions in messenger RNA during either transcription or translation. The scheme proposed for phenotypic mutation at the level of translation appears in Figure 17-3. This scheme predicts that certain base pair substitution mutants should be subject to phenotypic reversion by 5-fluorouracil; the prediction was confirmed by using T4rII mutants (Benzer and Champe, 1961). Champe and Benzer (1962b) reasoned that if the reversal occurred during translation, then it should only occur for AT mutants, and in particular only for those mutants in which the adenine residue occupied the transcribed strand of DNA. They observed that phenotypic reversion succeeded only with those mutants not susceptible to genotypic reversion by hydroxylamine, about a third of which responded to 5-fluorouracil.

Garen and Siddiqi (1962) demonstrated phenotypic reversion of *E. coli* alkaline phosphatase mutants by 5-fluorouracil, and observed that nonsense mutants responded preferentially. The majority of responding T4rII mutants also turned out to contain amber mutations. However, at least a few missense mutants in both systems do respond. It seems likely that a codon which is

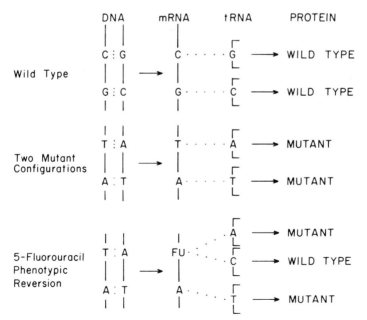

FIGURE 17-3. *Phenotypic reversion by 5-fluorouracil. Only one of the two possible types of mutant messenger RNA produced by G:C → A:T transitions can be affected.*

translated slowly would be particularly responsive to phenotypic reversion by 5-fluorouracil (Rosen, 1965), since a longer interval would exist during which the analogue could mispair with guanine residues in anticodons. Edlin (1965) studied the phenotypic reversion of amber mutations in many different cistrons of bacteriophage T4, and observed that early and late functions were easily distinguished by their time-dependent responses. He also observed that the effect was reversed by an excess of uracil. This result is consistent with the scheme presented in Figure 17-3, whereas if phenotypic reversion occurred during transcription instead of during translation, 5-fluorouracil would have to substitute for, and would be reversed by an excess of, cytosine.

Phenotypic suppression of T4rII mutants has also been achieved by using hydroxylamine (Levisohn, personal communication). Susceptibility is confined to mutants which are not genotypically reverted by hydroxylamine, indicating the involvement of A:T base pairs. The affected sites appear to be missense codons, since they are nonsuppressible and are frequently leaky. Prior treatment of cells before infection is not sufficient to obtain phenotypic suppression with hydroxylamine. These results suggest either that special codons lacking guanine and cytosine are affected by means of an unusual reaction with hydroxylamine (see Chapters 10 and 13), or that a special species of transfer RNA synthesized after virus infection is affected, perhaps by a reaction involving an anticodon cytosine residue.

Both 5-fluorouracil and 8-azaguanine have been reported to produce phenotypic reversion in *Neurospora* (Barnett and Brockman, 1962). Leaky mutants responded particularly well, suggesting the involvement of missense codons. The mutants usually responded either to both analogues, or to neither. It is possible in this system that 5-fluorouracil is acting through messenger RNA, and that 8-azaguanine is acting through transfer RNA (either during transcription or during translation).

Evidence has recently been obtained from a model *in vitro* system confirming the feasibility of modifying translational specificity by altering an anticodon. When glycine transfer RNA is treated with nitrous acid, a small fraction of the molecules subsequently insert glycine into a glutamic acid codon. This substitution is expected from the corresponding *anticodon* modification UCC → UUC produced by the oxidation deamination of cytosine (Carbon and Curry, 1968).

Phenotypic reversion is also efficiently promoted in *E. coli* by streptomycin and related antibiotics (Gorini and Kataja, 1964; Lederberg, Cavalli-Sforza, and Lederberg, 1964). Since the streptomycin effect can be abolished (but never enhanced) by mutations which modify cellular resistance to the drug, and since mutations affecting cellular resistance generally alter the ribosomes, it is likely that streptomycin modifies the translation process, perhaps by modifying the specificity of codon-anticodon interactions which occur on the ribosome (Couturier, Desmet, and Thomas, 1964; Gorini and Kataja, 1965; Anderson, Gorini, and Breckenridge, 1965; Gartner and Orias, 1966; Apirion and

Schlessinger, 1967). Streptomycin induces misreading of synthetic messengers *in vitro* by increasing, for instance, the probability that poly-U will direct the incorporation of leucine, serine, and isoleucine (Davies, Gilbert, and Gorini, 1964; Pestika, Marshall, and Nirenberg, 1965). Davies, Jones, and Khorana (1966) concluded that misreading is usually confined to pyrimidine residues, which may be misread as the other pyrimidine when they occupy the first position in a codon, and even as a purine when they occupy an internal position. In addition, misreading was observed to be strongly influenced by neighboring bases. Misreading *in vitro* appears to occur by two different mechanisms, one of which exhibits a requirement for single-stranded DNA of nonspecific origin as a cofactor but not as a messenger (Likover and Kurland, 1967).

The codon specificity of phenotypic mutation by streptomycin seems to include both missense and nonsense. Bissell (1965) reported that streptomycin caused the synthesis of abnormal β-galactosidase, and Couturier, Desmet, and Thomas (1964) reported the suppression of a temperature-sensitive mutant. Whitfield, Martin, and Ames (1966) reported suppression of one third of the missense mutants in *Salmonella*. Amber and ochre mutants are efficiently suppressed in both viral and bacterial systems (Valentine and Zinder, 1964; Orias and Gartner, 1966; Whitfield, Martin, and Ames, 1966; Scafati, 1967).

The complexity of the interaction between ribosome and transfer RNA is demonstrated by the observation that suppression by an altered transfer RNA is sharply inhibited by many mutations to streptomycin resistance (Otsuji and Aono, 1968); this inhibition is reversed by mutations to spectinomycin resistance in another gene (Kuwano, Endo, and Ohnishi, 1969).

Phenotypic reversion of diverse missense mutations has been induced in yeast by high osmotic pressure and also by pH alterations (Hawthorne and Friis, 1964; Bassel and Douglas, 1968; Loprieno et al., 1969). The results suggested that translational errors are not likely to be induced osmotically, and that tertiary or quaternary protein structure (polypeptide-folding or oligomerization respectively) is affected instead.

18. Complementation and polarity

A complementation test is performed by introducing two chromosomes or chromosome fragments, each of which is mutationally defective, into a common cytoplasm. If function (either growth or specific gene function) is increased compared to either single-chromosome control, then complementation has occurred. The complementation test, especially as it is applied to viruses, has already been introduced in Chapter 3, where it was employed in the definition of the gene itself. As we shall see, two rather distinct types of complementation occur, namely, between mutations in the same gene, or in different genes. Complementation has been extensively reviewed by Fincham (1966), and only its general principles will be reviewed here, together with a few comments about complementation in viral systems. The Fincham monograph should be consulted for representative data, and for discussions of some of the more subtle technical and conceptual difficulties which arise in the analysis of complementation. Finally, a warning to beginners: *complementation and recombination are distinct and unrelated phenomena*, although the tests for each are sometimes superficially similar.

When a mutational lesion in a gene not only inactivates that gene, but also partially or completely inactivates an adjacent gene or genes, the mutation is said to be polar. Polarity does not require that the mutational lesion extend physically into the adjacent gene (for instance by deletion). Inactivation of a gene *distant* from a mutated gene, although frequently observed, is not considered to result from polarity, for reasons which will become clear when the details of the phenomenon are discussed.

Both complementation and polarity are allele-specific phenomena: their occurrence depends upon the location and molecular configuration of the mutational site, but not necessarily or altogether upon the specific genes involved. Both phenomena arise because of effects of the mutational lesion upon translation of messenger RNA; they are discussed here because of their strong dependence upon the nature of the mutational lesion itself.

COMPLEMENTATION

Complementation tests are particularly easy with viral systems. They consist of mixed infections between pairs of mutants under nonpermissive conditions; the resulting burst sizes are compared with the burst sizes from single-infection controls. Consider a hypothetical screening experiment performed upon a large collection of conditional lethal mutants of two types, amber and temperature-sensitive. Initially, all the amber mutants are tested pairwise for complementation among themselves, while all the temperature-sensitive mutants are tested separately. The observed burst sizes are then arranged in two frequency distributions, as illustrated in Figure 18-1. The temperature-sensitive mutants produce a bimodal distribution, while the amber mutants produce a unimodal distribution. The nature of the bimodal distribution can be deduced from mapping experiments: the strongly complementing pairs of mutants will not often be closely linked, and will thus usually represent two different genes; the weakly complementing pairs of mutants will usually exhibit close linkage, and will thus usually represent the same gene. These tests, therefore, distinguish intergenic (intercistronic) and intragenic (intracistronic) complementation. (The term "interallelic complementation" is also frequently used to denote intragenic complementation.) It is clear from Figure 18-1, however, that phenotypic measurements alone cannot always reliably distinguish the two types of complementation, particularly for pairs of mutants falling into the saddle region of the bimodal distribution. In bacteriophage T4, for instance, intragenic

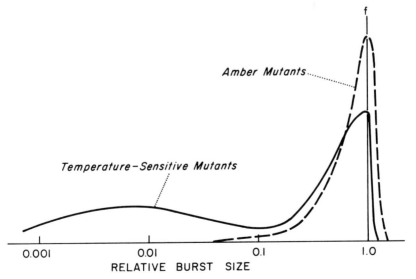

FIGURE 18-1.   *Hypothetical frequency distribution of relative burst sizes in viral complementation tests between randomly chosen pairs of mutants.*

complementation sometimes produces normal burst sizes (Bernstein, Edgar, and Denhardt, 1965).

The mechanism of intergenic complementation is obvious, and it is only necessary to consider the instances in which it fails to occur. There are several ways in which this may happen. First, one of the mutants in the test may be partially or wholly dominant, so that even mixed infection between the wild type and the mutant results in a sharply reduced burst size. Strongly dominant mutations are infrequent in microorganisms, but are occasionally observed. In bacteriophage T4, for example, the *r*II frameshift mutation *r*FC238 appears to cause the production of a small, toxic polypeptide (Barnett et al., 1967). Dominance will also result when a mutant polypeptide inactivates a mixed oligomer containing wild type polypeptides ("negative complementation"). A second type of intergenic noncomplementation can result whenever a requirement exists for joint synthesis and assembly, within a restricted space, of the proteins encoded by the two genes. A possible example may exist in the *hisC* region of *S. typhimurium*: an enzyme is produced which is composed of two nonidentical polypeptides. While *his* C is seperable into left and right halves by the polarity characteristics of its nonsense mutations, it is not divisible by complementation tests (Martin et al., 1966*b*). A third reason for the failure of intergenic complementation is polarity: if a polar mutation in gene A inactivates the adjacent gene B, then a complementation test between the A^- mutant and a second, B^- mutant will fail.

Intragenic complementation is a complex phenomenon; it may even occur among three mutant genes where the pairwise combinations all fail to complement (Inge-Vechtomov and Pavlenko, 1969). Its interpretation ultimately requires some understanding of the tertiary configuration of mutant proteins. Its occurrence is generally limited to missense mutants, especially leaky ones. It is observable in many genes, but not in all; in particular, it is most readily observed in genes whose proteins form aggregates or oligomers. Attempts to depict patterns of intragenic complementation and to correlate them with recombination maps, especially using data involving moderately large numbers of sites within a gene, have produced "complementation maps" of extremely complex and largely uninterpretable geometry.

A general theory of intragenic complementation was offered by Crick and Orgel (1964) and has been somewhat generalized by Fincham (1966). It assumes that intracistronic complementation usually occurs when the active form of the enzyme contains several identical subunits, and when oligomerization brings about a conformational change in the polypeptide. Mutant subunits either fail to form the oligomer, or else form an inactive version. A mixed oligomer, containing two different types of mutant subunits, may regain some activity if the wild type configuration at a position in one of the subunits can overcome the mutant configuration at the identical position in another subunit, particularly if this process can occur reciprocally. An entirely hypothetical and very simple example is given in Figure 18-2. A critical assumption in the theory of Crick and Orgel is that complementation patterns reflect the symmetry axes of

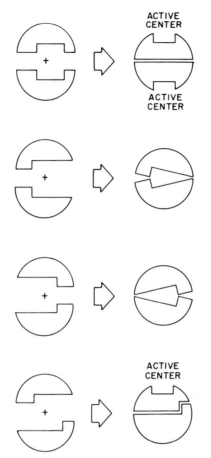

FIGURE 18-2. *Intragenic complementation dependent upon oligomerization of protein subunits. At the top, formation of a dyad has induced a conformational change activating both subunits. In the middle, highly imperfect dyads form from identical mutant subunits, but enzyme activation fails to occur. At the bottom, an imperfect dyad forms from two different mutant subunits, and one of the subunits is activated.*

the polypeptides within an oligomer. The interpretation of complex complementation maps may therefore require their rearrangement around symmetry axes (Gillie, 1966 and 1968).

A model has also been proposed for intragenic complementation which retains the Crick and Orgel symmetry requirements, but which supposes that enzymes may often possess active sites composed of amino acids in different polypeptide chains (McGavin, 1968); the active sites may be either catalytic or regulatory. This model generates quite complicated complementation maps using as few as two active sites.

Intragenic complementation has been demonstrated *in vitro* in a number of systems by polypeptide mixing experiments (see Fincham, 1966).

An exception has recently been reported to the rule that intragenic complementation is confined to missense mutations (Ullmann, Jacob, and Monod, 1967, 1968). Certain deletion and chain-termination mutants produce polypeptides representing the opposite ends of the *E. coli* β-galactosidase molecule. These peptide fragments can complement each other, both *in vivo* and *in vitro*.

The orderly sequence of events observed in the course of virus infection can sometimes complicate measurements of intragenic complementation, particularly when the mutations are located in a "late" gene. If a cell is mixedly infected with a pair of mutants whose mutational lesions are located toward opposite ends of a rather long gene, and if the gene functions rather late in the latent period, then wild type recombinant versions of the gene may be constructed by the time they are needed. This possibility is particularly prevalent in the T-even bacteriophages because of their high rates of genetic recombination (see the scale to Figure 4-3). At present there is no general and rigorous method to distinguish between recombination and complementation under such conditions.

When an equivocal response is obtained in complementation tests (for instance, a value corresponding to the saddle region in Figure 18-1), other methods must be used to define the genes containing the mutations under study. If both nonsense and missense mutations are readily available in the region, then mapping experiments, plus the assumption that nonsense mutations will not exhibit intragenic complementation, will usually unravel the situation (Figure 18-3). A pair of weakly complementing missense mutations which are bounded by a pair of noncomplementing nonsense mutations clearly fall within the same gene. Conversely, a pair of weakly complementing mutations which themselves

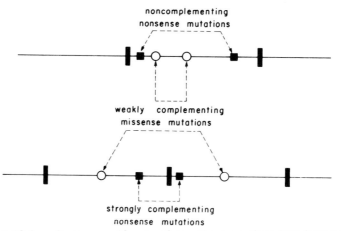

FIGURE 18-3.   *Sorting weakly complementing mutations into intragenic and intergenic categories by comparison with nonsense mutations and by mapping experiments. The heavy vertical bars represent the approximate locations of the ends of the cistrons.*

span a pair of strongly complementing nonsense mutations clearly fall into different genes.

POLARITY

Polarity is confined by definition to the limits of an operon. An operon is a group of contiguous cistrons under coordinate control (Jacob and Monod, 1961), and is transcribed into a polycistronic messenger RNA (Ohtaka and Spiegelman, 1963). Near the beginning of the operon is an operator region which interacts with the product of a regulatory gene; the regulatory gene, however, may map at a considerable distance from the operon.

Only nonsense mutants (chain-terminating base pair substitutions or frame-shift mutants) exhibit polarity. The gene harboring the mutational lesion is, of course, totally inactivated. The activities of all the genes in the operon which lie operator-distal to the mutational lesion are reduced in parallel by a common percentage, which may range from 0 to nearly 100. A gradient of polarity is frequently observed according to the location of the mutation within the gene, lesions at the operator-proximal end being most strongly polar (Newton et al., 1965; Baurle and Margolin, 1966a, b; Martin et al., 1966a; Yanofsky and Ito, 1966; Imamoto, Ito, and Yanofsky, 1967; Fink and Martin, 1967). The gradient may, however, be interrupted by distinct irregularities which correspond to sites within the cistron at which illegitimate initiation of translation may occur (Sarabhai and Brenner, 1967; Michels and Zipser, 1969; Newton, 1969). The degree of polarity can be reduced by deletions which shorten the interval between the mutational site and the end of the cistron (Newton, 1966; Zipser and Newton, 1967; Balbinder et al., 1968). When two nonsense mutations occupy the same gene, the resulting level of polarity is mostly but not exclusively determined by the first (operator-proximal) lesion (Yanofsky and Ito, 1966; Jordan and Saedler, 1967; Michels and Zipser, 1969; Newton, 1969).

Polarity gradients may differ strikingly, depending upon the gene, the operon, and the type of mutational lesion involved. Steep gradients tend to be observed in the first gene in an operon, while shallow gradients or no gradients at all appear in more distal genes (Fink and Martin, 1967). Certain mutations in the *E. coli* lactose operon produce extreme polarity, even when they fall at the operator-distal end of the gene (Malamy, 1966). Similar mutations in the *E. coli* galactose operon exhibit extreme polarity and no gradient, but are not susceptible to reversion analysis: although reverting spontaneously, they do not respond to base analogues, alkylating agents, or ICR-191 (Jordan, Saedler, and Starlinger, 1967). This type of mutation contains an extensive addition of nucleotide pairs (Jordan, Saedler, and Starlinger, 1968; Shapiro, 1969).

Polar mutations may also affect operator-proximal genes, depending upon the operon. In the *E. coli* tryptophan operon, the operator-proximal genes closest to the mutated gene are partially inactivated (antipolarity: Yanofsky and Ito, 1967), whereas in the galactose operon the operator-proximal genes exhibit increased activities (Jordan and Saedler, 1967). In *S. typhimurium,* operator-proximal genes in the histidine operon are not affected at all (Fink and Martin, 1967), but those in the tryptophan operon may be affected (Balbinder et al., 1968).

Nonsense mutations have variable effects upon levels of operon-specific messenger RNA. In bacteriophage T4, *r*II amber mutations do not affect the level of *r*II messenter RNA (Bautz, 1966). The level of *E. coli* lactose operon messenger RNA is very sharply reduced by strongly polar mutations (Attardi et al., 1963), but is restored by the addition of chloramphenicol (Contesse and Gros, 1968), which interrupts the usually strong coupling between protein synthesis and RNA synthesis. The amount of tryptophan operon messenger RNA is frequently depressed in the presence of strongly polar mutations (Imamoto and Yanofsky, 1967a, b). The portion of the messenger RNA operator-distal to the mutational lesion is specifically reduced in amount, probably because of degradation of mRNA from the 5' end (Morikawa and Imamoto, 1969; Morse et al., 1969).

When amber mutations are suppressed, their polarity is also suppressed (Newton et al., 1965; Yanofsky and Ito, 1966), although sometimes only partially (Martin et al., 1966a). Jordan and Saedler (1967) reported that the residual polarity observed after suppression no longer exhibited a gradient, but their data are in fact compatible with a shallow gradient. When nonpolar derivatives were obtained from strongly polar frameshift mutations, but no requirement was imposed for restoration of the primarily mutated gene, the "revertants" frequently contained a nonsense suppressor (Martin et al., 1966b), indicating that the polarity of frameshift mutations depends upon the chain-terminating codons which are generated distal to the mutational lesion (see Chapter 12).

Polarity has been demonstrated during the *in vitro* translation of single-stranded RNA bacteriophages (Engelhardt, Webster, and Zinder, 1967; Capecchi, 1967; Roberts and Gussin, 1967). Amber mutations in the viral coat protein inhibit the synthesis of viral RNA polymerase. A strong gradient is observed: an amber mutation affecting the sixth amino acid residue from the amino end of the coat protein is strongly polar, while amber mutations affecting positions 50, 54, 70, and 129 are at most weakly polar. The strongly polar amber mutation inhibits the formation of polysomes (polycistronic messenger RNA molecules carrying many ribosomes and producing many polypeptide chains), and also inhibits the formation of one or more viral-specific proteins (including the RNA polymerase).

The *in vitro* studies clearly establish that polarity is an aberration of translation. It is still necessary, however, to explain the mechanism of polarity, and in particular to explain why abnormal messenger RNA molecules are produced, why polarity depends upon the location of the lesion within the gene, and why certain aspects of polarity are so variable from system to system. Most of the suggested mechanisms postulate that ribosomes become functionally dissociated from messenger RNA after chain termination, unless chain initiation follows immediately, and that the probability of successful reinitiation at the next cistron decreases with the length of the untranslated interval (Imamoto, Ito, and Yanofsky, 1966; Malamy, 1966; Martin et al., 1966b). It is not at all clear, however, how the ribosome behaves in the untranslated interval: it may functionally dissociate by irreversibly falling off, or by continuing down the molecule but failing to reinitiate at the next cistron (perhaps because it has fallen out of the correct reading frame).

Schemes have been devised to explain some aspects of polarity in terms of the "texture" of the traversed but untranslated region. In particular, the steepness of the polarity gradient, the comparative behavior of base pair substitution and frameshift mutations, and the efficacy of suppression in removing the polar effect might all depend upon the frequency and distribution of specific codons in the untranslated region which are postulated to affect the behavior of passing ribosomes. The degree of polarity can also be supposed to depend upon the efficiency of chain initiation at the next cistron, which could vary according to the particular initiation signal employed. Voll (1967), in fact, has discovered polar mutations which do not inactivate the gene in which they occur, and which may be modifiers of initiation.

Another factor which would be expected to affect polarity is the coupling between translation and transcription postulated by Stent (1966). If ribosomes must attach to nascent messenger RNA molecules in order to promote transcription, then it is easy to imagine why polarity could result in the production of aborted messenger RNA molecules. The steepness of the polarity gradient might depend upon how close to the beginning of the polycistronic messenger RNA the mutated gene fell, since coupling might be strongest for the early portion of the messenger RNA molecule. Finally, if nonfunctioning ribosomes accumulated in the untranslated interval, they might eventually mechanically inhibit the translation of *preceding* cistrons (antipolarity).

19. Pseudomutation

Mutations are commonly defined as abrupt, heritable changes in an organism, and have been discussed thus far almost entirely in terms of altered base pair sequences. The problem of distinguishing between genic and extragenic alterations, however, can be exceedingly difficult, both conceptually and technically; an excellent discussion of this problem has been presented by Stadler (1954). This chapter will discuss some examples, both real and imaginary, of mutation-like events which nevertheless differ so strikingly in their underlying mechanisms from the processes which have been described in earlier pages, that it is necessary to place them in a special category. The philosophically inclined reader who wonders whether it is appropriate to call these mutations "false" may wish to begin with the question of whether recombinants are mutants.

One of the most peculiar examples of pseudomutation thus far described appeared precisely at the point when studies on mutation were initiated. The observations of de Vries (1901) on variation in *Oenothera,* the evening primrose, led him to found a general theory of mutation and its role in evolution. Only much later, however, did it become clear that the variants studied by de Vries are not mutations as we are now accustomed to use the word, but are instead the result of a strange, nearly unique, and very complex pattern of chromosome pairing, recombination, and segregation (see Muller, 1917, and Cleland, 1949). The additional examples of pseudomutation which will be considered here include events which arose as unrecognized examples of well-known processes such as genetic recombination, episomal transmission, and enzymatic modification of nucleic acids. The frequent occurrence of such phenomena should serve as a warning to students of mutation to keep their wits intact when faced with new and unusual data.

EPISOMAL PSEUDOMUTATION

An episome is a genetic element which may exist in the cell in either of two forms, integrated into the cellular chromosome, or as an autonomous, self-replicating cytoplasmic entity (Jacob and Wollman, 1961). The most common examples are temperate bacteriophages and bacterial sex and drug-resistance factors. Phenotypic changes commonly occur in cells upon both the entry and the departure of an episome. These changes can easily be misinterpreted as mutations if the investigator fails to recognize that an episome has arrived or departed, or if the change is not obviously or reproducibly related to an episome.

The integration of episomes into host chromosomes frequently results in the repression of most of the episomal genes, especially in the case of the temperate viruses. Not all viral functions are inactivated, however, and as a result, the infected cell usually acquires some new characteristics. One example which will already be familiar is the inhibition of T4rII replication which ensues when *E. coli* is lysogenized by bacteriophage λ. Phenomena of this type, collectively known as lysogenic conversion, can extend to a considerable range of characters, including, for instance, replacement of cellular antigens (Iseki and Sakai, 1953) and production of specific toxins (Freeman, 1951).

Most temperate bacteriophages integrate at only one or a few well-defined sites on the host chromosome, presumably regions containing base pair sequence homology with the viral chromosome. One remarkable bacteriophage, Mu 1, appears to violate this rule, Several percent of the cells newly lysogenized by this virus acquire auxotrophic mutations, the affected characters being diverse and unrelated (Taylor, 1963). The virus therefore appears to insert itself indiscriminately at many different sites, including sites within cistrons. The prophage is in fact located at or very near to the mutational site.

Putative examples of episomal-mediated mutations have been described for a number of other organisms, although the mechanisms in these systems are still not at all well understood. Some examples from bacterial systems have already been described in the section on mutator genes in Chapter 15. Baumiller (1967) has reported that an inherited virus of *Drosophila* is mutagenic when introduced into an uninfected stock, doubling the frequency of recessive lethal mutations. McClintock (1965) has proposed that a complicated series of replicating instabilities in maize are attributable to an episome-like element. In another maize system, a type of directed mutation ("paramutation") occurs in which a specific allele mutates in a particular manner with an extremely high frequency whenever it resides in the same cytoplasm with certain other alleles of the same gene (Brink, Styles, and Axtell, 1968).

Many temperate bacteriophages can transport fragments of the host genome from one cell to another, probably as an accident of maturation. This phenomenon, called transduction (Zinder, 1953), differs from episomal transmission since the transported fragment cannot replicate until it becomes integrated. Transduction can clearly mimic mutation, however, especially if the existence of the corresponding bacteriophage is not suspected (for instance, because both the donor and the recipient cell are lysogenic for the same temperate bacteriophage).

HOST-CONTROLLED VARIATION

Most examples of bacterial mutants resistant to viruses, and of viral mutants exhibiting extended host range, result from genetically determined modifications of receptor molecules. Occasionally, however, virus host range systems appear which clearly do not involve modified receptor systems, but are instead characterized by a set of rules like those illustrated in Table 19-1. The first two lines of the table are compatible with a conventional mutational alteration to extended host range, but the third line reveals that the "mutant" stock can revert completely to the "wild type" in the course of a single growth cycle. Other characteristics of the system also set the observed alterations apart from ordinary host range mutations: plating efficiencies on S-2 vary greatly according to the physiological state of the cells, and the variants of V-1 which are selected on S-2 arise with a random, rather than an exponential, distribution of clone sizes. The phenomenon is therefore commonly called host-controlled variation (or host-induced modification), and has been most recently reviewed by Arber (1965). At least three quite different mechanisms are known to operate, one being rather widely observed among bacteria and those bacteriophages whose DNA does not contain unusual components, and the other two being confined to particular viral systems.

TABLE 19-1

HOST-INDUCED MODIFICATION OF BACTERIOPHAGE PLATING EFFICIENCY*

Virus, with previous hosts indicated	Plating efficiency on bacterial strains	
	S-1	S-2
V-1	1	0.0001
V-2	1	1
V-2-1	1	0.0001

*V-1 denotes a virus stock grown on S-1 cells; V-2-1 denotes a stock grown on S-2 cells, and then passaged once on S-1 cells (from which a normal burst size is obtained).

In the most common type of host-controlled variation, the inability of a virus to multiply in a restrictive host such as S-2 results from the enzymatic degradation of the viral DNA. The chemical nature of the modification which protects the viral chromosome from attack is unproven, but is strongly suspected to be methylation. Physiological conditions which restrict the availability of methyl donors also restrict the ability of a cell to modify viral DNA. The restricting nuclease has been isolated; surprisingly, it requires as a cofactor the common methyl doner S-adenosylmethionine (Meselson and Yuan, 1968). The methylation of nucleic acids has been reviewed by Borek and Srinivasan (1966), and is known to occur both on the base and in sugar residues. Despite vigorous searches, quantitative differences in methylation frequencies have not yet been correlated with host-controlled variation. It is therefore likely that if methylation actually is involved, it is the qualitative distribution of methylated residues which is important. That being the case, it is not surprising that very complex patterns of host-controlled variation exist among several different strains of bacteria. It should also be noted that modification and restriction extend to DNA which is transferred between bacteria by sexual, transductional, and transformational mechanisms.

The DNA of the T-even bacteriophages is unique in possessing glucosylated 5HMC in place of cytosine. When grown in *Shigella dysenteriae,* or in mutants of *E. coli* unable to support glucosylation of the viral DNA, the resulting unglucosylated particles plate with very low efficiencies on wild type strains of *E. coli.* The DNA of the great majority of infecting particles is rapidly degraded. Mutants of bacteriophage T4 lacking the glucosylating enzymes (Georgopoulos, 1968), and mutants of *E. coli* unable to degrade unglucosylated T4 DNA (Revel, 1967), are both available.

Myxoviruses, including the influenza viruses, are single-stranded RNA viruses infecting a variety of birds and mammals. Host-controlled variation in Newcastle disease virus, which infects chickens, does not involve changes in plating efficiencies, but does involve changes in sensitivity to inactivation by heat, acid, and ultraviolet irradiation (Drake and Lay, 1962). Myxoviruses incorporate considerable amounts of preformed cellular macromolecules into the mature virus particle, and it is very likely that cell-specific differences in these components produce corresponding differences in the physical properties of the virus particles.

CRYPTIC RECOMBINATION

Spontaneous viral mutants usually arise during intracellular growth, and therefore during an interval when interactions may occur between the viral genome and any other genetic elements which may be present within the host cell. If any of the other genetic elements contain regions of homology with the virus, recombination between the two may occur. Such recombination, when resulting in the appearance of progeny virus particles carrying new markers, would be indistinguishable from ordinary mutation unless the previous existence within the host cell of that specific mutated gene could be established. Since bacteria frequently contain defective prophages (mutant prophages which are unable to complete a cycle of lytic multiplication), and since annealing experiments reveal considerable amounts of base pair sequence homology between temperate bacteriophages and their hosts (Cowie and McCarthy, 1963), cryptic recombination may be expected to occur rather frequently. This possibility has already been introduced in Chapter 14 as a source of certain viral mutants induced by ultraviolet irradiation, particularly by irradiation of the host cell; it will be recalled that irradiation is generally strongly recombinogenic, as well as mutagenic.

Mutations of bacteriophage T3 sometimes arise under conditions which strongly suggest the presence of a recombinational mechanism (Fraser, 1957). Although T3 is a lytic virus, it employs no unusual DNA components, and some aspects of its growth suggest that it may be semi-temperate in character. Certain T3 mutants characteristically arise within special cells with greatly extended latent periods, during which they elongate into filamentous "snakes." The burst from a single snake may contain mutations at several different sites, and the frequency of mutation is strongly increased by ultraviolet irradiation of the host cell. The specific mutants which appear tend to be characteristic of the particular host cell employed. Repeated serial isolations of mutants using a given host strain may eventually produce stable T3 derivatives which no longer "mutate" in that strain. It is clearly easy to interpret these mutants as arising from recombination between T3 and a related but defective prophage carried in many strains of *E. coli*.

A completely different type of cryptic recombination may be imagined in which a pair of homologous chromosomes are known to be present, but in which they do not appear to carry allelic differences. Suppose that both chromosomes have serine codons at a particular position, but that the two codons are different (UCU and AGU, for instance, which might have arisen from a common ACU [threonine] ancestor). Intracodon recombination could then produce both threonine and cysteine (UGU) codons, with corresponding changes in the phenotype. What would appear to be a mutational event would in fact merely be a process which revealed the existence of mutations that actually arose far in the past.

GENE CONVERSION

During meiosis in certain fungal systems, the two parental chromosomes duplicate shortly before recombination normally takes place. As a result, four strands are present during the period when crossing over occurs, and these are ultimately distributed in four ordered spores (Figure 19-1). Recombination between well-separated markers is typically reciprocal, but recombination between a pair of very close markers frequently produces an anomaly called gene conversion. When gene conversion occurs, the input into a cross consists of a 2:2 ratio of alleles at a given site, but the output consists of a 3:1 ratio. Conversion occurs much more frequently than does ordinary mutation of the allele. Furthermore, the converted allele is not altered in a random manner, but instead changes to the other allele in the cross. Gene conversion therefore resembles a directed mutational process.

The mechanism of gene conversion is poorly understood. Its close association with recombination has led to proposals that DNA degradation and repair are involved (Holliday, 1964; Whitehouse and Hastings, 1965). The partly hypothetical mechanism of recombination outlined in Figure 3-4 suggests how this might happen: strand degradation in a hybrid region could erase a marker, and repair

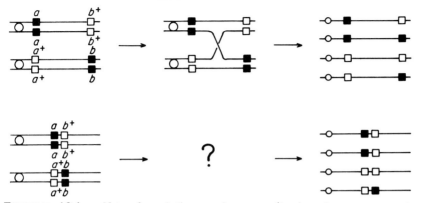

FIGURE 19-1.   *Normal meiotic crossing over (top) and gene conversion (bottom).*

synthesis employing a strand from the other chromosome as a template could then replace it with the other marker. The complete process is probably more complex than this, however, and appears to involve coordinated degradation and repair in different molecules; note that each of the chromosomes shown in Figure 19-1 is probably composed of a double-stranded DNA molecule.

SELF-PRIMED GENES

Consider the hypothetical example of a gene whose continuing activity depends upon the presence of the gene product: if the supply of gene product is temporarily depleted, either by physiological accident or by design, the gene is inactivated. If at a later time the gene product or an acceptable substitute is reintroduced, the gene is reactivated. Although no chromosomal alteration has occurred, the gene appears to have mutated and then to have reverted. In terms of information content, of course, the gene product itself is a component of such a gene, and its loss corresponds to a deletion mutation. No clear-cut example of such a mutation has been described yet, but many of the genes which determine structural components required for transcription and translation (such as RNA polymerase, amino acid activating enzymes, and ribosomes) could clearly become involved in such a mutation.

The concept of self-priming genes is an old one; it was invoked, for example, to explain the behavior of the killer trait in *Paramecium* (Sonneborn, 1943), where a nuclear gene is required for the continued presence of a cytoplasmic factor, but the cytoplasmic factor is not produced until some is already present. (The cytoplasmic factor later turned out to be a self-replicating entity.) A bacterial system has been described which also exhibits self-priming activity (Novick and Weiner, 1957). The enzymes of the *E. coli* lactose operon are only induced by a threshold concentration of inducer. In the presence of sub-threshold concentrations, an occasional cell does become induced, perhaps because of a random local accumulation of inducer molecules. Once induced, the cell produces a permease which actively concentrates the inducer, thus maintaining the cell in the induced state. If a physiological accident were to temporarily make the cell unable to concentrate the inducer, however, it could revert to the uninduced state.

NONGENIC INHERITANCE

Most of the examples of pseudomutation cited in this chapter concern events which, although unusual, nevertheless affect the constitution and behavior of genes which are composed of nucleic acids. It would hardly be fair to discuss unusual modes of mutation, however, without pointing out a nearly unique example of inheritance mediated by a genetic apparatus composed mainly of sand (Jennings, 1937). The rhizopod protozoan *Difflugia corona* is encased in a shell composed of a chitin-like matrix in which are embedded large numbers of extremely minute siliceous particles. The shell also contains an oral orifice which is surrounded by a ring of "teeth." *Difflugia* replicates by growth and division of the internal mass, one naked daughter cell being extruded though the mouth. This daughter cell rapidly secretes her own shell, using her sister's teeth as a template for her own. If teeth are broken off, either by accident or by design, then the descendents of a cell continue to be deficient in teeth in the corresponding location. Over the course of many generations, however, the mouth gradually returns to its original configuration. The inheritance of the mouth parts can therefore undergo not only spontaneous and induced mutation, but also reversion.

A number of other examples of morphological modifications in protozoa, surgically induced and thereafter inherited, have been discussed by Sonneborn (1964). These observations clearly indicate that genetic information can be stored in molecules other than nucleic acids, thus reinforcing the significance of the central dogma of biology: only cells (and not nucleic acids) make cells. At least for the time being.

Bibliography

Abe, M., and Tomizawa, J., Replication of the *Escherichia coli* K12 chromosome, *Proc. Natl. Acad. Sci. U.S.*, 58:1911–1918, 1967.

Adams, M. H., The genotypically and phenotypically heat resistant forms in the T5 species of bacteriophage, *Ann. Inst. Pasteur*, 84:164–174, 1953.

Adelberg, E. A., Mandel, M., and Chen, G. C. C., Optimal conditions for mutagenesis by N-methyl-N'-nitro-N-nitrosoguanidine in *Escherichia coli*, *Biochem. Biophys. Res. Commun.*, 18:788–795, 1965.

Alderson, T., Chemically induced delayed germinal mutation in *Drosophila*, *Nature*, 207:164–167, 1965.

Allen, M. K., and Yanofsky, C., A biochemical and genetic study of reversion with the A-gene A-protein system of *Escherichia coli* tryptophan synthetase, *Genetics*, 48:1065–1083, 1963.

Ames, B. N., and Hartman, P. E., The histidine operon, *Cold Spring Harbor Symp. Quant. Biol.*, 28:349–356, 1963.

Ames, B. N., and Whitfield, H. J., Frameshift mutagenesis in *Salmonella*, *Cold Spring Harbor Symp. Quant. Biol.*, 31:221–225, 1966.

Anderson, T. F., The inheritance of requirements for adsorption cofactors in the bacterial virus T4, *J. Bacteriol.*, 55:651–658, 1948.

Anderson, T. F., Rappaport, C., and Muscatine, N. A., On the structure and osmotic properties of phage particles, *Ann. Inst. Pasteur*, 84:5–14, 1953.

Anderson, W. F., The effect of tRNA concentration on the rate of protein synthesis, *Proc. Natl. Acad. Sci. U.S.*, 62:566–573, 1969.

Anderson, W. F., Gorini, L., and Breckenridge, L., Role of ribosomes in streptomycin-activated suppression, *Proc. Natl. Acad. Sci. U.S.*, 54:1076–1083, 1965.

Apirion, D., Altered ribosomes in a suppressor strain of *Escherichia coli*, *J. Mol. Biol.*, 16:285–301, 1966.

Apirion, D., and Schlessinger, D., The loss of phenotypic suppression in streptomycin-resistant mutants, *Proc. Natl. Acad. Sci. U.S.*, 58:206–212, 1967.

Arber, W., Host-controlled modification of bacteriophages, *Ann. Rev. Microbiol.*, 19:365–378, 1965.

Ardashnikov, S. N., Soyfer, V. N., and Goldfarb, D. M., Induction of h-mutations in the extracellular phage T2 by γ-irradiation, *Biochem. Biophys. Res. Commun.*, 16:455–459, 1964.

Attardi, G., Naono, S., Rouvière, J., Jacob, F., and Gros, F., Production of messenger RNA and regulation of protein synthesis, *Cold Spring Harbor Symp. Quant. Biol.*, 28:363–372, 1963.

Auerbach, C., Spontaneous mutations in dry spores of *Neurospora crassa, Z. Vererbungslehre*, **90**:335–346, 1959.

Auerbach, C., *Mutation: An Introduction to Research on Mutagenesis. Part I: Methods*, Oliver and Boyd, Edinburgh and London, 1962.

Auerbach, C., Lethal sectoring and the origin of complete mutants in *Schizosaccharomyces pombe, Mutation Res.*, **4**:875–878, 1967.

Auerbach, C., and Robson, J. M., Chemical production of mutations, *Nature*, **157**:302, 1946.

Auerbach, C., and Robson, J. M., The production of mutations by chemical substances, *Proc. Roy. Soc. Edinburgh (B)*, **62**:271–283, 1947.

Avery, O. T., MacLeod, C. M., and McCarty, M., Studies on the chemical nature of the substance inducing transformation of pneumococcal types, *J. Exp. Med.*, **79**:137–158, 1944.

Bacon, D. F., and Treffers, H. P., Spontaneous and mutator-induced reversions of an *Escherichia coli* auxotroph, *J. Bacteriol.*, **81**:786–793, 1961.

Baker, R., and Tessman, I., The circular genetic map of phage S13, *Proc. Natl. Acad. Sci. U.S.*, **58**:1438–1445, 1967.

Baker, R., and Tessman, I., Different mutagenic specificities in phages S13 and T4: *in vivo* treatment with N-methyl-N'-nitro-N-nitrosoguanidine, *J. Mol. Biol*, **35**:439–448, 1968.

Balbinder, E., Blume, A. J., Weber, A., and Tamaki, H., Polar and antipolar mutants in the tryptophan operon of *Salmonella typhimurium, J. Bacteriol*, **95**:2217–2229, 1968.

Ballbinder, E., The fine structure of the loci *tryC* and *tryD* of *Salmonella typhimurium*, 2. Studies of reversion patterns and the behavior of specific alleles during recombination, *Genetics*, **47**:545–559, 1962.

Ball, C., and Roper, J. A., Studies on the inhibition and mutation of *Aspergillus nidulans* by acridines, *Genetical Res.*, **7**:207–221, 1966.

Barnett, L., Brenner, S., Crick, F. H. C., Shulman, R. G., and Watts-Tobin, R. J., Phase-shift and other mutants in the first part of the *r*II B cistron of bacteriophage T4, *Phil. Trans. Roy. Soc. London (B)*, **252**:487–560, 1967.

Barnett, W. E., and Brockman, H. E., Induced phenotypic reversion by 8-azaguanine and 5-fluorouracil, *Biochem. Biophys. Res. Commun.*, **7**:199–203, 1962.

Barricelli, N. A., A note on the elementary theory of mutant clone-size distribution in the progeny of phages treated extracellularly with a mutagen, *Virology*, **27**:630–633, 1965.

Barricelli, N. A., and Del Zoppo, G., Genotypic reversion by methylene blue: the orientation of guanine-hydroxymethylcytosine at mutated sites in *r*II mutants of phage T4, *Molec. Gen. Genetics*, **101**:51–58, 1968.

Baricelli, N. A., and Womack, F., A radiation-genetic method to decide whether one or both of the DNA strands injected by a virus into its host transmit the hereditary information to the progeny, *Virology*, **27**:589–599, 1965.

Bassel, J., and Douglas, H. C., Osmotic remedial response in a galactose-negative mutant of *Saccharomyces cerevisiae, J. Bacteriol.*, **95**:1103–1110, 1968.

Bauerle, R. H., and Margolin, P., The functional organization of the tryptophan gene cluster in *Salmonella typhimurium, Proc. Natl. Acad. Sci. U.S.*, **56**:111–118, 1966*a*.

Bauerle, R. H., and Margolin, P., A multifunctional enzyme complex in the tryptophan pathway of *Salmonella typhimurium*: comparison of polarity and pseudopolarity mutations, *Cold Spring Harbor Symp. Quant. Biol.*, **31**:203–214, 1966*b*.

Baumiller, R. C., Virus induced point mutation, *Nature*, **214**:806–807, 1967.

Bautz, E. K. F., The effect of a nonsense triplet on the stability of messenger RNA, *J. Mol. Biol.*, **17**:298–301, 1966.

Bautz, E., and Freese, E., On the mutagenic effect of alkylating agents, *Proc. Natl. Acad. Sci. U.S.*, **46**:1585–1594, 1960.

Bautz, F. A., and Bautz, E. K. F., Mapping of deletions in a non-essential region of the phage T4 genome, *J. Mol. Biol.*, **28**:345–355, 1967.

Bautz-Freese, E., and Freese, E., Induction of reverse mutations and cross reactivation of nitrous acid-treated phage T4, *Virology*, **13**:19–30, 1961.

Baylor, M. B., Hessler, A. Y., and Baird, J. P., The circular linkage map of bacteriophage T2H, *Genetics*, **51**:351–361, 1965.

Baylor, M. B., Hurst, D. D., Allen, S. L., and Bertani, E. T., The frequency and distribution of loci affecting host range in the coliphage T2H, *Genetics*, **42**:104–120, 1957.

Becker, E. F., Zimmerman, B. K., and Geiduschek, E. P., Structure and function of cross-linked DNA, 1. Reversible denaturation and *Bacillus subtilis* transformation, *J. Mol. Biol.*, **8**:377–391, 1964.

Beckwith, J., A deletion analysis of the *lac* operator region in *Escherichia coli*, *J. Mol. Biol.*, **8**:427–430, 1964.

Beckwith, J., Signer, E., and Epstein, W., Transposition of the lac region of *E. coli*, *Cold Spring Harbor Symp. Quant. Biol.*, **31**:393–401, 1966.

Bellin, J. S., and Grossman, L. I., Photodynamic degradation of nucleic acids, *Photochem. Photobiol.* **4**:45–53, 1965.

Belser, W. L., Mutational studies in *Serratia marinorubra*, *Science*, **134**:1427, 1961.

Bendigkeit, H. E., Cell size, DNA synthesis, and mutation in chemostats, *Bact. Proc.*, 102, 1966.

Benzer, S., Fine structure of a genetic region in bacteriophage, *Proc. Natl. Acad. Sci. U.S.*, **41**:344–354, 1955.

Benzer, S., The elementary units of heredity, in *The Chemical Basis of Heredity*, edited by W. D. McElroy and B. Glass, pp. 70–93, The Johns Hopkins Press, Baltimore, 1957.

Benzer, S., On the topology of the genetic fine structure, *Proc. Natl. Acad. Sci. U.S.*, **45**:1607–1620, 1959.

Benzer, S., On the topography of the genetic fine structure. *Proc. Natl. Acad. Sci. U.S.*, **47**:403–415, 1961*a*.

Benzer, S., Genetic fine structure, *Harvey Lectures*, **56**:1–21, 1961*b*.

Benzer, S., Behavioral mutants of *Drosophila* isolated by countercurrent distribution, *Proc. Natl. Acad. Sci. U.S.*, **58**:1112–1119, 1967.

Benzer, S., and Champe, S. P., Ambivalent *r*II mutants of phage T4, *Proc. Natl. Acad. Sci. U.S.*, **47**:1025–1038, 1961.

Benzer, S., and Champe, S. P., A change from nonsense to sense in the genetic code, *Proc. Natl. Acad. Sci. U.S.*, **48**:1114–1121, 1962.

Benzer, S., and Freese, E., Induction of specific mutations with 5-bromouracil, *Proc. Natl. Acad. Sci. U.S.*, **44**:112–119, 1958.

Berens, S., and Shugar, D., Ultraviolet absorption spectra and structure of halogenated uracils and their glycosides, *Acta Biochem. Polon.*, **10**:25–48, 1963.

Berg, C. M., and Curtiss, R., Transposition derivatives of an Hfr strain of *Escherichia coli* K-12, *Genetics*, **56**:503–525, 1967.

Berg, P., Fancher, H., and Chamberlain, M., The synthesis of mixed poly-nucleotides containing ribo- and deoxyribonucleotides by purified preparations of DNA polymerase from *Escherichia coli*, in *Informational Macro-molecules*, edited by H. J. Vogel, V. Bryson, and J. O. Lempen, pp. 467–483, Academic Press, New York, 1963.

Berger, H., Genetic analysis of T4D phage heterozygotes produced in the presence of 5-fluorodeoxyuridine, *Genetics*, **52**: 729–746, 1965.

Berger, H., Brammar, W. J., and Yanofsky, C, Spontaneous and ICR191-A-induced frameshift mutations in the *A* gene of *Escherichia coli* tryptophan synthetase, *J. Bacteriol*, **96**:1672–1679, 1966.

Berger, H., and Yanofsky, C., Suppressor selection for amino acid replacements expected on the basis of the genetic code, *Science*, **156**:394–397, 1967.

Bernstein, H., The effect on recombination of mutational defects in the DNA-polymerase and deoxycytidylate hydroxymethylase of phage T4D, *Genetics*, **56**:755–769, 1967.

Bernstein, H., Edgar, R. S., and Denhardt, G. H., Intragenic complementation among temperature sensitive mutants of bacteriophage T4D, *Genetics*, **51**:987–1002, 1965.

Bertani, L. E., Host-dependent induction of phage mutants and lysogenization, *Virology*, **12**:553–569, 1960.

Bessman, M. J., Lehman, I. R., Adler, J., Zimmerman, S. B., Simms, E. S., and Kornberg, A., Enzymatic synthesis of deoxyribonucleic acid, 3. The incorporation of pyrimidine and purine analogues into deoxyribonucleic acid, *Proc. Natl. Acad. Sci. U.S.*, **44**:633–640, 1958.

Beukers, R., The effect of proflavin on U.V.-induced dimerization of thymine in DNA, *Photochem. Photobiol.*, **4**:935–937, 1965.

Beukers, R., and Berends, W., The effect of paramagnetic substances on the conversion of some pyrimidines by ultraviolet irradiation, *Biochim. Biophys. Acta*, **38**:573–575, 1960*a*.

Beukers, R., and Berends, W., Isolation and identification of the irradiation product of thymine, *Biochim. Biophys. Acta*, **41**:550–551, 1960*b*.

Bissell, D. M., Formation of an altered enzyme by *Escherichia coli* in the presence of neomycin, *J. Mol. Biol.*, **14**:619–622, 1965.

Bode, W., Zur Natur der Ausgedehnten Mutationen des Coliphagen T4, *Z. Vererbungslehre*, **94**:190–199, 1963.

Bode, W., Lysis inhibition in *Escherichia coli* infected with bacteriophage T4, *J. Virology*, **1**:948–955, 1967.

Böhme, H., Streptomycin-abhängige Mutanten von *Proteus mirabilis* und ihre Verwendurg in Mutationsversuchen mit Manganchlorid, *Biol. Zentralbl.*, **80**:5–32, 1961*a*.

Böhme, H., Über Ruckmutationen und Suppressormutationen bei *Proteus mirabilis*, *Z. Vererbungslehre*, **92**:197–204, 1961*b*.

Böhme, H., Genetic instability of an ultraviolet-sensitive mutant of *Proteus mirabilis*, *Biochem. Biophys. Res. Commun.*, **28**:191–196, 1967.

Böhme, H., and Geissler, E., Repair of lesions induced by photodynamic action and by ethyl methanesulfonate in *E. coli*, *Molec. Gen. Genetics*, **103**:228–232, 1968.

Böhme, H., and Wacker, A., Mutagenic activity of thiopyronine and methyleneblue in combination with visible light, *Biochem. Biophys. Res. Commun.*, **12**:137–139, 1963.

Borek, E., and Srinivasan, P. R., The methylation of nucleic acids, *Ann. Rev. Biochem.*, **35**:275–298, 1966.

Botstein, D., and Jones, E. W., Nonrandom mutagenesis of the *Escherichia coli* genome by nitrosoguanidine, *J. Bacteriol.*, **98**:847–848, 1969.

Boyce, R. P., and Howard-Flanders, P., Release of ultraviolet light-induced thymine dimers from DNA in *E. coli* K-12, *Proc. Natl. Acad. Sci. U.S.*, **51**:293–300, 1964.

Bradley, T. B., Wohl, R. C., and Rieder, R. F., Hemoglobin Gun Hill: deletion of five amino acid residues and impaired heme-globin binding, *Science,* **157**:1581–1583, 1967.

Brammar, W. J., Berger, H., and Yanofsky, C., Altered amino acid sequences produced by reversion of frameshift mutants of tryptophan synthetase A gene of *E. coli, Proc. Natl. Acad. Sci. U.S.,* **58**:1499–1506, 1967.

Brendel, M., Induction of mutation in phage T4 by extracellular treatment with methylene blue and visible light, *Molec. Gen. Genetics,* **101**:111-115, 1968.

Brendel, M., and Kaplan, R. W., Photodynamische Mutationsauslösung und Inaktivierung beim *Serratia*–Phagen X durch Methylenblau und Licht, *Molec. Gen. Genetics,* **99**:181–190, 1967.

Brenner, S., Physiological aspects of bacteriophage genetics, *Adv. Virus Res.,* **6**:137–158, 1959.

Brenner, S., and Barnett, L., Genetic and chemical studies on the head protein of bacteriophage T2 and T4, *Brookhaven Symp. Biol.,* **12**:86–93, 1959.

Brenner, S., Barnett, L., Crick, F. H. C., and Orgel, A., The theory of mutagenesis, *J. Mol. Biol.,* **3**:121–124, 1961.

Brenner, S., Barnett, L., Katz, E. R., and Crick, F. H. C., UGA: a third nonsense triplet in the genetic code, *Nature,* **213**:449–450, 1967.,

Brenner, S., Benzer, S., and Barnett, L., Distribution of proflavin-induced mutations in the genetic fine structure, *Nature* **182**:983–985, 1958.

Brenner, S., and Stretton, A. O. W., Phase shifting of *amber* and *ochre* mutants, *J. Mol. Biol.,* **13**:944–946, 1965.

Brenner, S., Stretton, A. O. W., and Kaplan, S., Genetic code: the 'nonsense' triplets for chain termination and their suppression, *Nature,* **206**:994–998, 1965.

Bresch, C., Genetical studies on bacteriophage T1, *Ann. Inst. Pasteur,* **84**:157, 1953.

Bresch, C., Zum Paarungsmechanismus von Bakteriophagen, *Z. Naturforsch.,* **10B**:545–561, 1955.

Bresler, S. E., Kreneva, R. A., and Kushev, V. V., Correction of molecular heterozygotes in the course of transformation, *Molec. Gen. Genetics,* **102**:257–268, 1968.

Bridges, B. A., Dennis, R. E., and Munson, R. J., Differential induction and repair of ultraviolet damage leading to true reversions and external suppressor mutations of an ochre codon in *Escherichia coli* B/r WP2, *Genetics,* **57**:897–908, 1967.

Bridges, B. A., Law, J., and Munson, R. J., Mutagenesis in *Escherichia coli,* 2. Evidence for a common pathway for mutagenesis by ultraviolet light, ionizing radiation and thymine deprivation, *Molec. Gen. Genetics,* **103**:266–273, 1968.

Bridges, B. A., and Munson, R. J., Mutagenesis in *Escherichia coli*: evidence for the mechanism of base change mutation by ultraviolet radiation in a strain deficient in excision-repair, *Proc. Roy. Soc. London (B),* **171**:213–226, 1968.

Brink, R. A., Styles, E. D., and Axtell, J. D., Paramutation: directed genetic change, *Science,* **159**:161–170, 1968.

Britten, R. J., and Kohne, D. E., Repeated sequences in DNA, *Science,* **161**:529–540, 1968.

Brockman, H. E., and de Serres, F. J., Induction of *ad-3* mutants of *Neurospora crassa* by 2-aminopurine, *Genetics,* **48**:597–604, 1963.

Brockman, H. E., and Goben, W., Mutagenicity of a monofunctional alkylating agent derivative of acridine in *Neurospora, Science,* **147**:750–751, 1965.

Brody, S., and Yanofsky, C., Suppressor gene alteration of protein primary structure, *Proc. Natl. Acad. Sci. U.S.,* **50**:9–16, 1963.

Brookes, P., and Lawley, P. D., Effects of alkylating agents on T2 and T4 bacteriophages, *Biochem. J.*, **89**:138–144, 1963.

Brown, D. F., X-ray-induced mutation in extracellular bacteriophage T4, *Mutation Res.*, 3:365–373, 1966a.

Brown, D. F., X-ray-induced mutations in extracellular phages, *Nature*, 212:1595–1596, 1966b.

Brown, D. M., Hewlins, M. J. E., and Schell, P., The tautomeric state of *N*(4)-hydroxy- and of *N*(4)-amino-cytosine derivatives, *J. Chem. Soc. C*, 1925–1929, 1968.

Brown, D. M., McNaught, A. D., and Schell, P., The chemical basis of hydrazine mutagenesis, *Biochem. Biophys. Res. Commun.*, 24:967–971, 1966.

Brown, D. M., and Schell, P., The reaction of hydroxylamine with cytosine and related compounds, *J. Mol. Biol.*, 3:709–710, 1961.

Burton, K., Lunt, M. R., Petersen, G. B., and Siebke, J. C., Studies of nucleotide sequences in deoxyribonucleic acid, *Cold Spring Harbor Symp. Quant. Biol.*, 28:27–34, 1963.

Bussard, A., Naono, S., Gros, F., and Monod, J., Effets d'un analogue de l'uracile sur les propriétés d'une protéine enzymatique synthétisée en sa présence, *C. R. Acad. Sci.*, 250:4049–4051, 1960.

Cairns, J., The application of autoradiography to the study of DNA viruses, *Cold Spring Harbor Symp. Quant. Biol.*, 27:311–318, 1962.

Cairns, J., The bacterial chromosome and its manner of replication as seen by autoradiography, *J. Mol. Biol.*, 6:208–213, 1963.

Cairns, J., Stent, G. S., and Watson, J. D., *Phage and the Origins of Molecular Biology*, Cold Spring Harbor Laboratory of Quantitative Biology, Cold Spring Harbor, New York, 1966.

Calberg-Bacq, C. M., Delmelle, M., and Duchesne, J., Inactivation and mutagenesis due to the photodynamic action of acridines and related dyes on extracellular bacteriophage T4B, *Mutation Res.*, 6:15–24, 1968.

Campbell, A., Ordering of genetic sites in bacteriophage λ by the use of galactose-transducing defective phages, *Virology*, 9:293–305, 1959.

Campbell, A., Sensitive mutants of bacteriophage λ, *Virology*, 14: 22–32, 1961.

Campbell, A. M., Episomes, *Adv. Genetics*, 11:101–145, 1962.

Capecchi, M. R., Polarity *in vitro*, *J. Mol. Biol.*, 30:213–217, 1967.

Capecchi, M. R., and Gussin, G. N., Suppression *in vitro*: identification of a serine-sRNA as a "nonsense" suppressor, *Science*, 149:417–422, 1965.

Carbon, J., Berg, P., and Yanofsky, C., Studies of missense suppression of the tryptophan synthetase A-protein mutant A36, *Proc. Natl. Acad. Sci. U.S.*, 56:764–771, 1966.

Carbon, J., and Curry, J. B., A change in the specificity of transfer RNA after partial deamination with nitrous acid, *Proc. Natl. Acad. Sci. U.S.*, 59:467–474, 1968.

Carlson, E. A., *The Gene: A Critical History*, W. B. Saunders, Philadelphia, 1966.

Carlson, E. A., Sederoff, R., and Cogan, M., Evidence favoring a frameshift mechanism for ICR-170 induced mutations in *Drosophila melanogaster*, *Genetics*, 55:295–313, 1967.

Cavilla, C. A., and Johns, H. E., Inactivation and photoreactivation of the T-even phages as a function of the inactivating ultraviolet wavelength, *Virology*, 24:349–358, 1964.

Cerdá-Olmedo, E., and Hanawalt, P. C., Macromolecular action of nitrosoguanidine in *Escherichia coli*, *Biochim. Biophys. Acta.*, 142:450–464, 1967.

Cerdá-Olmedo, E., and Hanawalt, P. C., Diazomethane as the active agent in nitrosoguanidine mutagenisis and lethality, *Molec. Gen. Genetics*, 101:191–202, 1968.

Cerdá-Olmedo, E., Hanawalt, P. C., and Guerola, N., Mutagenesis of the replication point by nitrosoguanidine: map and pattern of replication of the *Escherichia coli* chromosome, *J. Mol. Biol.,* **33**:705–719, 1968.

Champe, S. P., and Benzer, S., An active cistron fragment. *J. Mol. Biol.,* **4**:288–292, 1962*a*.

Champe, S. P., and Benzer, S., Reversal of mutant phenotypes by 5-fluorouracil: an approach to nucleotide sequences in messenger-RNA, *Proc. Natl. Acad. Sci. U.S.,* **48**:532–546, 1962*b*.

Chan, L. M., and Van Winkle, Q., Interaction of acriflavin with DNA and RNA, *J. Mol. Biol.,* **40**:491–495, 1969.

Chernik, T. P., and Krivisky, A. S., Mutagenesis in the course of the genetic transformation of *Bacillus subtilis,* and the effect thereon of the roentgen irradiation of the transforming DNA, *Genetics (USSR),* **4** no. 6:75–86, 1968.

Christensen, J. R., and Saul, S. H., A cold-sensitive mutant of bacteriophage T1, *Virology,* **29**:497–499, 1966.

Clarke, C. H., Caffeine- and amino acid-effects upon try$^+$ revertant yield in U.V.-irradiated hcr$^+$ and hcr$^-$ mutants of *E. coli* B/r, *Molec. Gen. Genetics,* **99**:97–108, 1967.

Cleland, R. E., A botanical nonconformist, *Sci. Monthly,* **68**:35–41, 1949.

Cockayne, E. A., *Inherited Abnormalities of the Skin and its Appendages,* Oxford University Press, Oxford, 1933.

Cohen, S. N., and Hurwitz, J., Transcription of complementary strands of phage λ DNA *in vivo* and *in vitro, Proc. Natl. Acad. Sci. U.S.,* **57**:1759–1766, 1967.

Cohen, S. S., Flaks, J. G., Barner, H. D., Loeb, M. R., and Lichtenstein, J., The mode of action of 5-fluorouracil and its derivatives, *Proc. Natl. Acad. U.S.,* **44**:1004–1012, 1958.

Contesse, G., and Gros, F., Action du chloramphenicol sur la transcription de l'opéron lactose chez des mutants polaires d'*Escherichia coli, C. R. Acad. Sci.,* **266**:262–265. 1968.

Cook, A., and Lederberg, J., Recombination studies of lactose nonfermenting mutants of *Escherichia coli* K-12, *Genetics,* **47**:1335–1353, 1962.

Cooper, P. D., The mutation of poliovirus by 5-fluorouracil, *Virology,* **22**:186–192, 1964.

Coughlin, C. A., and Adelberg, E. A., Bacterial mutation by thymine starvation, *Nature,* **178**:531–532, 1956.

Couturier, M., Desmet, L., and Thomas, R., High pleiotropy of streptomycin mutations in *Escherichia coli, Biochem. Biophys. Res. Commun.,* **16**:244–248, 1964.

Cowie, D. B., and McCarthy, B. J., Homology between bacteriophage λ DNA and *E. coli* DNA, *Proc. Natl. Acad. Sci. U.S.,* **50**:537–543, 1963.

Cox, E. C., and Yanofsky, C., Altered base ratios in the DNA of an *Escherichia coli* mutator strain, *Proc. Natl. Acad. Sci. U.S.,* **58**:1895–1902, 1967.

Crick, F. H. C., Codon-anticodon pairing: the wobble hypothesis, *J. Mol. Biol.,* **19**:548–555, 1966.

Crick, F. H. C., Barnett, L., Brenner, S., and Watts-Tobin, R. J., General nature of the genetic code for proteins, *Nature,* **192**:1227–1232, 1961.

Crick, F. H. C., and Orgel, L. E., The theory of inter-allelic complementation, *J. Mol. Biol.,* **8**:161–165, 1964.

Crick, F. H. C., and Watson, J. D., A structure for deoxyribose nucleic acid, *Nature,* **171**:737, 1953.

Cummings, D. J., Sedimentation and biological properties of T-phages of *Escherichia coli, Virology,* **23**:408–418, 1964.

Davern, C. I., The inhibition and mutagenesis of an RNA bacteriophage by 5-fluorouracil, *Australian J. Biol. Sci.,* **17**:726–737, 1964.

Davidson, J. N., Leslie, I., and White, J. C., The nucleic-acid content of the cell, *Lancet,* **260**:1287–1290, 1951.

Davies, J., Gilbert, W., and Gorini, L., Streptomycin, suppression, and the code, *Proc. Natl. Acad. Sci. U.S.,* **51**:883–890, 1964.

Davies, J., Jones, D. S., and Khorana, H. G., A further study of misreading of codons induced by streptomycin and neomycin using ribopolynucleotides containing two nucleotides in alternating sequence as templates, *J. Mol. Biol.,* **18**:48–57, 1966.

Davis, R. W., and Davidson, N., Electron-microscopic visualization of deletion mutations, *Proc. Natl. Acad. Sci. U.S.,* **60**:243–250, 1968.

Dawson, G. W. P., and Smith-Keary, P. F., Episomic control of mutation in *Salmonella typhimurium, Heredity,* **18**:1–20, 1963.

de Groot, B., The bar properties, in particular glucosylation of deoxyribonucleic acid, in crosses of bacteriophages T2 and T4, *Genetical Res.,* **9**:149–158, 1967.

de Jong, W. W. W., Went, L. N., and Bernini, L. F., Haemoglobin Leiden: deletion of $\beta 6$ or 7 glutamic acid, *Nature,* **220**:788–790, 1968.

Delbrück, M., Biochemical mutants of bacterial viruses, *J. Bacteriol.,* **56**:1–16, 1948.

Dellweg, H., and Oprée, W., Die photodynamische Wirkung von Thiopyronin auf Nucleinsäuren, *Biophysik,* **3**:241–248, 1966.

DeMars, R. I., Chemical mutagenesis in bacteriophage T2, *Nature,* **172**:964, 1953.

DeMars, R. I., Luria, S. E., Fisher, H., and Levinthal, C., The production of incomplete bacteriophage particles by the action of proflavin and the properties of the incomplete particles, *Ann. Inst. Pasteur,* **84**:113–128, 1953.

Demerec, M., Frequency of deletions among spontaneous and induced mutations in *Salmonella, Proc. Natl. Acad. Sci. U.S.,* **46**:1075–1079, 1960.

Demerec, M., "Selfers"–attributed to unequal crossovers in *Salmonella, Proc. Natl. Acad. Sci. U.S.,* **48**:1696–1704, 1962.

Demerec, M., Gillespie, D. H., and Mizobuchi, K., Genetic structure of the *cysC* region of the *Salmonella* genome, *Genetics,* **48**:997–1009, 1963.

Demerec, M., and Hanson, J., Mutagenic action of manganous chloride, *Cold Spring Harbor Symp. Quant. Biol.,* **16**:215–227, 1951.

Demerec, M., Lahr, E. L., Miyake, T., Galehran, I., Balbinder, E., Baric, S., Hashimoto, K., Glanville, E. V., and Gross, J. D., Bacterial genetics, *Carnegie Inst. Wash. Yearbook,* **370**:390–406, 1957.

Denhardt, D. T., and Silver, R. B., An analysis of the clone size distribution of ΦX174 mutants and recombinants, *Virology,* **30**:10–19, 1966.

Denhardt, D. T., and Sinsheimer, R. L., The process of infection with bacteriophage ΦX174, 3. Phage maturation and lysis after synchronized infection, *J. Mol. Biol.,* **12**:641–646, 1965a.

Denhardt, D. T., and Sinsheimer, R. L., The process of infection with bacteriophage ΦX174, 4. Replication of the viral DNA in a synchronized infection, *J. Mol. Biol.,* **12**:647–662, 1965b.

Denhardt, D. T., and Sinsheimer, R. L., The process of infection with bacteriophage ΦX174, 5. Inactivation of the phage-bacterium complex by decay of ^{32}P incorporated in the infecting particle, *J. Mol. Biol.,* **12**:663–673, 1965c.

Devoret, R., Influence du genotype de la bacterie hote sur la mutation du phage λ produite par le rayonnement ultraviolet, *C. R. Acad. Sci.,* **260**:1510–1513, 1965.

de Vries, H., *Die Mutationstheorie*, Vol. 1, Veit, Leipzig, 1901.

de Waard, A., Paul, A. V., and Lehman, I. R., The structural gene for deoxyribonucleic acid polymerase in bacteriophages T4 and T5, *Proc. Natl. Acad. Sci. U.S.*, 54:1241–1248, 1965.

Dobzhansky, T., *Mankind Evolving*, Yale University Press, New Haven, 1962.

Doerfler, W., and Hogness, D. S., Gene orientation in bacteriophage lambda as determined from the genetic activities of heteroduplex DNA formed *in vitro*, *J. Mol. Biol.*, 33:661–678, 1968.

Doermann, A. H., Lysis and lysis inhibition with *Escherichia coli* bacteriophage, *J. Bacteriol.*, 55:257–276, 1948.

Doermann, A. H., and Boehner, L., An experimental analysis of bacteriophage T4 heterozygotes, 1. Mottled plaques from crosses involving six rII loci, *Virology*, 21:551–567, 1963.

Doermann, A. H., and Boehner, L., An experimental analysis of bacteriophage T4 heterozygotes, 2. Distribution in a density gradient, *J. Mol. Biol.*, 10:212–222, 1964.

Doermann, A. H., and Parma, D. H., Recombination in bacteriophage T4, *J. Cell. Physiol.*, 70 Suppl. 1:147–164, 1967.

Donohue, J., Hydrogen-bonded helical configurations of polynucleotides, *Proc. Natl. Acad. Sci. U.S.*, 42:60–65, 1956.

Doneson, I. N., and Shankel, D. M., Mutational synergism between radiations and methylated purines in *Escherichia coli*, *J. Bacteriol.*, 87:61–67, 1964.

Doudney, C. O., and Haas, F. L., Modification of ultraviolet induced mutation frequency and survival in bacteria by post-irradiation treatment, *Proc. Natl. Acad. Sci. U.S.*, 44:390–401, 1958.

Doudney, C. O., and Haas, F. L., Gene replication and mutation induction in bacteria, *J. Mol. Biol.*, 1:81–83, 1959.

Doudney, C. O., and Haas, F. L., Some biochemical aspects of the post-irradiation modification of mutation response to ultraviolet light in bacteria, *Bact. Proc.*, 69, 1960.

Doudney, C. O., White, B. F., and Bruce, B. J., Acriflavin modification of nucleic acid formation, mutation induction and survival in ultraviolet light exposed bacteria, *Biochem. Biophys. Res. Commun.*, 15:70–75, 1964.

Dove, W. F., The genetics of the lambdoid phages, *Ann. Rev. Genetics*, 2:305–340, 1968.

Dowell, C. E., Cold-sensitive mutants of bacteriophage ΦX174, 1. A mutant blocked in the eclipse function at low temperature, *Proc. Natl. Acad. Sci. U.S.*, 58:958–961, 1967.

Drake, J. W., Polyimines mutagenic for bacteriophage T4B, *Nature*, 197:4871, 1963a.

Drake, J. W., Properties of ultraviolet-induced rII mutants of bacteriophage T4, *J. Mol. Biol.*, 6:268–283, 1963b.

Drake, J. W., Mutational activation of a cistron fragment, *Genetics*, 48:767–773, 1963c.

Drake, J. W., Studies on the induction of mutations in bacteriophage T4 by ultraviolet irradiation and by proflavin, *J. Cell. Comp. Physiol.*, 64 Suppl. 1:19–31, 1964.

Drake, J. W., Heteroduplex heterozygotes in bacteriophage T4 involving mutations of various dimensions, *Proc. Natl. Acad. Sci. U.S.*, 55:506–512, 1966a.

Drake, J. W., Spontaneous mutations accumulating in bacteriophage T4 in the complete absence of DNA replication, *Proc. Natl. Acad. Sci. U.S.*, 55:738–743, 1966b.

Drake, J. W., Ultraviolet mutagenesis in bacteriophage T4, 1. Irradiation of extracellular phage particles, *J. Bacteriol.*, **91**:1775–1780, 1966c.

Drake, J. W., Ultraviolet mutagenesis in bacteriophage T4, 2. Photoreversal of mutational lesions, *J. Bacteriol.* **92**:144–147, 1966d.

Drake, J. W., The length of the homologous pairing region for genetic recombination in bacteriophage T4, *Proc. Natl. Acad. Sci. U.S.*, **58**:962–966, 1967.

Drake, J. W., Comparative spontaneous mutation rates, *Nature*, **221**:1128–1132, 1969.

Drake, J. W., and Allen, E. F., Antimutagenic DNA polymerases of bacteriophage T4, *Cold Spring Harbor Symp. Quant. Biol.*, **33**:339–344, 1968.

Drake, J. W., Allen, E. F., Forsberg, S. A., Preparata, R. M., and Greening. E. O., The genetic control of mutation rates in bacteriophage T4, *Nature*, **221**:1128–1132, 1969.

Drake, J. W., and Lay, P. A., Host-controlled variation in NDV, *Virology*, **17**:56–64, 1962.

Drake, J. W., and McGuire, J., Characteristics of mutations appearing spontaneously in extracellular particles of bacteriophage T4, *Genetics*, **55**:387–398, 1967a.

Drake, J. W., and McGuire, J., Properties of *r* mutants of bacteriophage T4 photodynamically induced in the presence of thiopyronin and psoralen, *J. Virology*, **1**:260–267, 1967b.

Drummond, D. S., Pritchard, N. J., Simpson-Gildemeister, V. F. W., and Peacocke, A. R., Interaction of aminoacridines with deoxyribonucleic acid: viscosity of the complexes, *Biopolymers*, **4**:971–987, 1966.

Drummond, D. S., Simpson-Gildemeister, V. F. W., and Peacocke, A. R., Interaction of aminoacridines with deoxyribonucleic acid: effects of ionic strength, denaturation, and structure, *Biopolymers*, **3**:135–153, 1965.

Dulbecco, R., Experiments on photoreactivation of bacteriophages inactivated with ultraviolet radiation, *J. Bacteriol.*, **59**:329–348, 1950.

Dulbecco, R., and Vogt, M., Studies on the induction of mutations in poliovirus by proflavin, *Virology*, **5**:236–243, 1958.

Dunn. L. C., *A Short History of Genetics*, McGraw-Hill, New York, 1965.

Edgar, R. S., Mapping experiments with *rII* and *h* mutants of bacteriophage T4D, *Virology*, **6**:215–225, 1958.

Edgar, R. S., and Epstein, R. H., Inactivation by ultraviolet light of an acriflavin-sensitive gene function in phage T4D, *Science*, **134**:327–328, 1961.

Edgar, R. S., Feynman, R. P., Klein, S., Lielausis, I., and Steinberg, C. M., Mapping experiments with *r* mutants of bacteriophage T4D, *Genetics*, **47**:179–186, 1962.

Edgar, R. S., and Lielausis, I., Temperature-sensitive mutants of bacteriophage T4D: their isolation and genetic characterization, *Genetics*, **49**:649–662, 1964.

Edgar, R. S., and Wood, W. B., Morphogenesis of bacteriophage T4 in extracts of mutant-infected cells, *Proc. Natl. Acad. Sci. U.S.*, **55**:498–505, 1966.

Edlin, G., Gene regulation during bacteriophage T4 development, 1. Phenotypic reversion of T4 amber mutants by 5-fluorouracil, *J. Mol. Biol.* **12**:363–374, 1965.

Eggers, H. J., and Tamm, I., Spectrum and characteristics of the virus inhibitory action of 2-(a-hydroxybenzyl)-benzimidazole, *J. Exp. Med.*, **113**:657–682, 1961.

Eisen, H. A., Fuerst, C. R., Siminovitch, L., Thomas, R., Lambert, L., Pereira da

Silva, L., and Jacob, F., Genetics and physiology of defective lysogeny in *E. coli* K12(λ): studies of early mutants, *Virology,* **30**:224–241, 1966.

Eisenstark, A., Eisenstark, R., and van Sickle, R., Mutation of *Salmonella typhimurium* by nitrosoguanidine, *Mutation Res.,* **2**:1–10, 1965.

Eisenstark, A., and Rosner, J. L., Chemically induced reversions in the *cysC* region of *Salmonella typhimurium, Genetics,* **49**:343–355, 1964.

Eisinger, J., and Shulman, R. G., The precurser of the thymine dimer in ice, *Proc. Natl. Acad. Sci. U.S.,* **58**:895–900, 1967.

Ellison, S. A., Feiner, R. R., and Hill, R. F., A host effect on bacteriophage survival after ultraviolet irradiation, *Virology,* **11**:294–296, 1960.

Ellmauer, H., and Kaplan, R. W., Auslosung von Klarplaque-Mutationen durch UV-Bestrahlung des freien Phagen κ von *Serratia marcescens, Naturwiss.,* **46**:150, 1959.

Emrich, J., Lysis of T4-infected bacteria in the absence of lysozme, *Virology,* **35**:158–165, 1968.

Engelhardt, D. L., Webster, R. E., Wilhelm, R. C., and Zinder, N. D., *In vitro* studies on the mechanism of suppression of a nonsense mutation, *Proc. Natl. Acad. Sci. U.S.,* **54**:1791–1797, 1965.

Engelhardt, D. L., Webster, R. E., and Zinder, N. D., Amber mutants and polarity *in vitro, J. Mol. Biol.,* **29**:45–58, 1967.

Englesberg, E., Anderson, R. L., Weinberg, R., Lee, N., Hoffee, P., Hutterhauer, G., and Boyer, H., L-arabinose-sensitive, L-ribulose 5-phosphate 4-epimerase-deficient mutants of *Escherichia coli, J. Bacteriol.,* **84**:137–146, 1962.

Epstein, R. H., Bolle, A., Steinberg, C. M., Kellenberger, E., Boy de la Tour, E., Chevalley, R., Edgar, R. S., Susman, M., Denhardt, G. H., and Lielausis, A., Physiological studies of conditional lethal mutants of bacteriophage T4D, *Cold Spring Harbor Symp. Quant. Biol.,* **28**:375–392, 1963.

Evans, N. A., Hidalgo-Salvatierra, O., and McLaren, A. D., Effect of HCN on the photochemistry of uracil, *Biophys. Soc. Abstracts,* 79, 1967.

Fenner, F., and Comben, B. M., Genetic studies with mammalian poxviruses, 1. Demonstration of recombination between two strains of vaccinia virus, *Virology,* **5**:530–548, 1958.

Fermi, G., and Stent, G. S., Effects of chloramphenicol and of multiplicity of infection on induced mutation in bacteriophage T4, *Z. Vererbungslehre,* **93**:177–187, 1962.

Fernandez, B., Haas, F. L., and Wyss, O., Induced host-range mutations in bacteriophage, *Proc. Natl. Acad. Sci. U.S.,* **39**:1052–1057, 1953.

Feughelman, M., Langridge, R., Seeds, W. E., Stokes, A. R., Wilson, H. R., Hooper, C. W., Wilkins, M. H. F., Barclay, R. K., and Hamilton, L. D., Molecular structure of deoxyribose nucleic acid and nucleoprotein, *Nature,* **175**:834–838, 1955.

Fincham, J. R. S., *Genetic Complementation,* W. A. Benjamin, Inc., New York, 1966.

Fink, G. R., Klopotowski, T., and Ames, B. N., Histidine regulatory mutants in *Salmonella typhimurium,* 4. A positive selection for polar histidine-requiring mutants from histidine operator constitutive mutants, *J. Mol. Biol.,* **30**:81–95, 1967.

Fink, G. R., and Martin, R. G., Translation and polarity in the histidine operon, 2. Polarity in the histidine operon, *J. Mol. Biol.,* **30**:97–107, 1967.

Folsome, C. E., Specificity of induction of T4*r*II mutants by ultraviolet irradiation of extracellular phages, *Genetics,* **47**:611–622, 1962.

Folsome, C. E., Unstable *r*II mutants of bacteriophage T4, *Biochem. Biophys. Res. Commun.,* **14**:156–160, 1964.

Folsome, C. E., and Levin, D., Detecting reversion in T4rII bacteriophage to r^+ induced by ultraviolet irradiation, *Nature*, **192**:1306, 1961.

Franklin, N., Extraordinary recombinational events in *Escherichia coli*, their independence of the *rec*$^+$ function, *Genetics*, **55**:699–707, 1967.

Fraser, D. K., Host range mutants and semitemperate mutants of bacteriophage T3, *Virology*, **3**:527–553, 1957.

Freeman, V. J., Studies on the virulence of bacteriophage-infected strains of *Corynebacterium diptheriae*, *J. Bacteriol.*, **61**:675–688, 1951.

Freese, E., The difference between spontaneous and base analogue induced mutation of phage T4, *Proc. Natl. Acad. Sci. U.S.*, **45**:622–633, 1959*a*.

Freese, E., The specific mutagenic effect of base analogues on Phage T4, *J. Mol. Biol.*, **1**:87–105, 1959*b*.

Freese, E., On the molecular explanation of spontaneous and induced mutations, *Brookhaven Symp. Biol.*, **12**:63–73, 1959*c*.

Freese, E. B., Transitions and transversions induced by depurinating agents, *Proc. Natl. Acad. Sci. U.S.*, **47**:540–545, 1961.

Freese, E., Molecular mechanism of mutations, in *Molecular Genetics, Part I*, edited by J. H. Taylor, pp. 207–269, Academic Press, New York, 1963.

Freese, E. B., The mutagenic effect of hydroxyaminopurine derivatives in phage T4, *Mutation Res.*, **5**:299–301, 1968.

Freese, E., Bautz, E., and Freese, E. B., The chemical and mutagenic specificity of hydroxylamine, *Proc. Natl. Acad. Sci. U.S.*, **47**:845–855, 1961.

Freese, E., Bautz-Freese, E., and Bautz, E., Hydroxylamine as a mutagenic and inactivating agent, *J. Mol. Biol.*, **3**:133–143, 1961.

Freese, E. B., and Freese, E., Two separable effects of hydroxylamine on transforming DNA, *Proc. Natl. Acad. Sci. U.S.*, **52**:1289–1297, 1964.

Freese, E., and Freese, E. B., The oxygen effect on deoxyribonucleic acid inactivation by hydroxylamine, *Biochem.*, **4**:2419–2433, 1965.

Freese, E. B., and Freese, E., Induction of pure mutant clones by repair of inactivating DNA alterations in phage T4, *Genetics*, **54**:1055–1067, 1966.

Freese, E. B., and Freese, E., On the specificity of DNA polymerase, *Proc. Natl. Acad. Sci. U.S.*, **57**:650–657, 1967.

Freese, E., Freese, E. B., and Graham, S., The oxygen-dependent reaction of hydroxylamine with nucleotides and DNA, *Biochim. Biophys. Acta*, **123**:17–25, 1966.

Freese, E., and Strack, H. B., Induction of mutations in transforming DNA by hydroxylamine, *Proc. Natl. Acad. Sci. U.S.*, **48**:1796–1803, 1962.

Freifelder, D., Davison, P. F., and Geiduschek, E. P., Damage by visible light to the acridine orange-DNA complex, *Biophys. J.*, **1**:389–400, 1961.

Freifelder, D., and Uretz, R. B., Mechanism of photoinactivation of coliphage T7 sensitized by acridine orange, *Virology*, **30**:97–103, 1966.

Fresco, J. R., and Alberts, B. M., The accommodation of noncomplementary bases in helical polyribonucleotides and deoxyribonucleic acids, *Proc. Natl. Acad. Sci. U.S.*, **46**:311–321, 1960.

Funatsu, G., and Fraenkel-Conrat, H., Location of amino acid exchanges in chemically evoked mutants of tobacco mosaic virus, *Biochem.*, **3**:1356–1362, 1964.

Garen, A., and Siddiqi, O., Suppression of mutations in the alkaline phosphatase structural cistrons of *E. coli*, *Proc. Natl. Acad. Sci. U.S.*, **48**:1121–1127, 1962

Gartner, T. K., and Orias, E., Effects of mutations to streptomycin resistance on the rate of translation of mutant genetic information, *J. Bacteriol.*, **91**:1021–1028, 1966.

Gefter, M., Hausmann, R., Gold, M., and Hurwitz, J., The enzymatic methylation of ribonucleic acid and deoxyribonucleic acid, 10. Bacteriophage T3-induced S-adenosylmethionine cleavage, *J. Biol. Chem.,* **241**:1995–2006, 1966.

Geiduschek, E. P., "Reversible" DNA, *Proc. Natl. Acad. Sci. U.S.,* **47**:950–955, 1961.

Georgopoulos, C. P., Location of glucosyl transferase genes on the genetic map of phage T4, *Virology,* **34**:364–366, 1968.

Gesteland, R. F., Salser, W., and Bolle, A., *In vitro* synthesis of T4 lysozyme by suppression of amber mutations, *Proc. Natl. Acad. Sci. U.S.,* **58**:2036–2042, 1967.

Gierer, A., and Schramm, G., Infectivity of ribonucleic acid from tobacco mosaic virus, *Nature,* **177**:702, 1956.

Gillie, O. J., The interpretation of complementation data, *Genetical Res.,* **8**:9–31, 1966.

Gillie, O. J., Interpretations of some large nonlinear complementation maps, *Genetics,* **58**:543–555, 1968.

Gilmore, R. A., Stewart, J. W., and Sherman, F., Amino acid replacements resulting from super-suppression of a nonsense mutant of yeast, *Biochim. Biophys. Acta,* **161**:270–272, 1968.

Gold, M., Hausmann, R., Maitra, U., and Hurwitz, J., The enzymatic methylation of RNA and DNA, 8. Effects of bacteriophage infection on the activity of the methylating enzymes, *Proc. Natl. Acad. Sci. U.S.,* **52**:292–297, 1964.

Goldberg, E. B., The amount of DNA between genetic markers in phage T4, *Proc. Natl. Acad. Sci. U.S.,* **56**:1457–1463, 1966.

Goldberg, I. H., and Rabinowitz, M., The incorporation of 5-ribosyluracil triphosphate into RNA in nuclear extracts of mammalian cells, *Biochem. Biophys. Res. Commun.,* **6**:394–398, 1961.

Goldfarb, D. M., Nesterova, G. F., and Kuznetsova, V. N., Localization of h^+ mutations of T phage, induced by various mutagens, *Soviet Genetics,* **2**:19–22, 1966.

Goldschmidt, E. P., Matney, T. S., and Bausum, H. T., Genetic analysis of mutations from streptomycin dependence to independence in *Salmonella typhimurium, Genetics,* **47**:1475–1487, 1962.

Goldstein, A., and Smoot, J. S., A strain of *Escherichia coli* with an unusually high rate of auxotrophic mutation, *J. Bacteriol.,* **70**:588–595, 1955.

Goodman, H. M., Abelson, J., Landy, A., Brenner, S., and Smith, J. D., Amber suppression: a nucleotide change in the anticodon of a tyrosine transfer RNA, *Nature,* **217**,:1019–1024, 1968.

Gorin, A. I., Spitkovsky, D. M., Tikchonenko, T. I., and Tseytlin, P. I., The secondary structure of phage DNA in phage particles, *Biochim. Biophys. Acta,* **134**:490–492, 1967.

Gorini, L., and Beckwith, J. R., Suppression, *Ann. Rev. Microbiol.,* **20**:401–422, 1966.

Gorini, L., and Kataja, E., Phenotypic repair by streptomycin of defective genotypes of *E. coli, Proc. Natl. Acad. Sci. U.S.,* **51**:487–493, 1964.

Gorini, L., and Kataja, E., Suppression activated by streptomycin and related antibiotics in drug sensitive strains, *Biochem. Biophys. Res. Commun.,* **18**:656–663, 1965.

Gottschling, H., and Freese, E., Incorporation of 2-aminopurine into the desoxyriboncleic acid of bacteria and bacteriophages, *Z. Naturforsch.,* **16B**:515–519, 1961.

Gottschling, H., and Heidelberger, C., Fluorinated pyrimidines, 19. Some biological effects of 5-trifluoromethyluracil and 5-trifluoromethyl-2'-deoxyuridine on *Escherichia coli* and bacteriophage T4B, *J. Mol. Biol.*, 7:541–560, 1963.

Goulian, M., Kornberg, A., and Sinsheimer, R. L., Enzymatic synthesis of DNA, 24. Synthesis of infectious phage ΦX174 DNA, *Proc. Natl. Acad. Sci. U.S.*, 58:2321–2328, 1967.

Granoff, A., Induction of Newcastle disease virus mutants with nitrous acid, *Virology*, 13:402–408, 1961.

Green, D. M., and Krieg, D. R., The delayed origin of mutants induced by exposure of extracellular phage T4 to ethyl methane sulfonate, *Proc. Natl. Acad. Sci. U.S.*, 47:65–72, 1961.

Greenstock, C. L., Brown, I. H., Hunt, J. W., and Johns, H. E., Photodimerization of pyrimidine nucleic acid derivatives in aqueous solution and the effect of oxygen, *Biochem. Biophys. Res. Comm.*, 4:431–436, 1967.

Greenstock, C. L., and Johns, H. E., Photosensitized dimerization of pyrimidines, *Biochem. Biophys. Res. Comm.*, 30:21–27, 1968.

Greer, S. B., Growth inhibitors and their antagonists as mutagens and antimutagens in *Escherichia coli*, *J. Gen. Microbiol.*, 18:543–564, 1958.

Greer, S., and Zamenhof, S., Studies on depurination of DNA by heat, *J. Mol. Biol.*, 4:123–141, 1962.

Grigg, G. W., and Sergeant, D., Compound loci and coincident back-mutation in *Neurospora*, *Z. Vererbungslehre*, 92:380–388, 1961.

Grigg, G. W., and Stuckey, J., The reversible suppression of stationary phase mutation in *Escherichia coli* by caffeine, *Genetics*, 53:823–834, 1966.

Gross, J. D., Karamata, D., and Hempstead, P. G., Temperature-sensitive mutants of *B. subtilis* defective in DNA synthesis, *Cold Spring Harbor Symp. Quant. Biol.*, 33:307–312, 1968.

Grossman, L., The effect of ultraviolet-irradiated polyuridylic acid in cell-free protein synthesis in *E. coli*, *Proc. Natl. Acad. Sci. U.S.*, 48:1609–1614, 1962.

Grossman, L., The effects of ultraviolet-irradiated polyuridylic acid in cell-free protein synthesis in *Escherichia coli*, 2. The influence of specific photoproducts, *Proc. Natl. Acad. Sci. U.S.*, 50:657–664, 1963.

Grossman, L., Studies on mutagenesis induced *in vitro*, *Photochem. Photobiol.*, 7:727–735, 1968.

Guerrini, F., and Fox, M. S., Effects of DNA repair in transformation-heterozygotes of pneumococcus, *Proc. Natl. Acad. Sci. U.S.*, 59:1116–1123, 1968*b*.

Guerrini, F., and Fox, M. S., Genetic heterozygosity in pneumococcal transformation, *Proc. Natl. Acad. Sci. U.S.*, 59:429–436, 1968*a*.

Guglielminetti, R., The role of lethal sectoring in the origin of complete mutations in *Schizosaccharomyces pombe*, *Mutation Res.*, 5:225–229, 1968.

Gundersen, W. B., New type of streptomycin resistance resulting from action of the episomelike mutator factor in *Escherichia coli* B(ORNL), *J. Bacteriol.*, 86:510–516, 1963.

Gundersen, W. B., Transduction of the mu-factor in *Escherichia coli*, *Acta Path. Microbiol. Scand.*, 65:621–626, 1965.

Gundersen, W. B., Jyssum, K., and Lie, S., Genetic instability with episome-mediated transfer in *Escherichia coli*, *J. Bacteriol.*, 83:616–623, 1962.

Günther, H. L., and Prusoff, W. H., Decrease in sensitivity to ultraviolet irradiation of *Streptococcus faecalis* grown in media supplemented with 6-azathymine, an analogue of thymine, *Biochim. Biophys. Acta*, 55:778–780, 1962.

Gupta, N. K., and Khorana, H. G., Missense suppression of the tryptophan synthetase A-protein mutant A78, *Proc. Natl. Acad. Sci. U.S.*, **56**:772–779, 1966.

Gurney, T., and Fox, M. S., Physical and genetic hybrids formed in bacterial transformation, *J. Mol. Biol.*, **32**:83–100, 1968.

Haefner, K., Concerning the mechanism of ultraviolet mutagenesis. A micro-manipulatory pedigree analysis in *Schizosaccharomyces pombe, Genetics*, **57**:169–178, 1967*a*.

Haefner, K., A remark to the origin of pure mutant clones observed after UV treatment of *Schizosaccharomyces pombe, Mutation Res.*, **4**:514–516, 1967*b*.

Hall, D. H., and Tessman, I., Linkage of T4 genes controlling a series of steps in pyrimidine biosynthesis, *Virology*, **31**:442–448, 1967.

Hall, Z. W., and Lehman, I. R., An *in vitro* transversion by a mutationally altered T4-induced DNA polymerase, *J. Mol. Biol.*, **36**:321–333, 1968.

Halle S., 5-Azacytidine as a mutagen for arboviruses. *J. Virology* **2**:1228–1229, 1968.

Hamers, R., and Hamers-Casterman, C., Synthesis by *Escherichia coli* of an abnormal β-galactosidase in the presence of thiouracil, *J. Mol. Biol.*, **3**:166–174, 1961.

Harm, W., Mutants of phage T4 with increased sensitivity to ultraviolet, *Virology*, **19**:66–71, 1963*a*.

Harm, W., in *Repair from Genetic Radiation* edited by Sobels, pp. 107–124, Pergamon Press, New York, 1963*b*.

Harm, W., On the control of UV-sensitivity of phage T4 by the gene *x. Mutation Res.*, **1**:344 354, 1964.

Harm, W., Comment on the relationship between UV reactivation and host-cell reactivation in phage, *Virology*, **29**:494, 1966*a*.

Harm, W., The role of host-cell repair in liquid-holding recovery of U.V.-irradiated *Escherichia coli, Photochem. Photobiol.*, **5**:747–760, 1966*b*.

Harm, W., Differential effect of acriflavin and caffeine on various ultraviolet-irradiated *Escherichia coli* strains and T1 phage, *Mutation Res.*, **4**:93–110, 1967.

Harm, W., and Hillebrandt, B., A nonphotoreactivable mutant of *E. coli* B, *Photochem. Photobiol,* **1**:271–272, 1962.

Harm, W., and Hillebrandt, B., Kompetitive Hemmung der Photo-Reaktivierung von uv-inaktivierten T4-Phagen durch stark uv-bestrahlte Phagen-DNA, *Z. Naturforsch.*, **18B**:294–300, 1963.

Haug, A., and Sauerbier, W., Influence of pH and oxygen on 315 mμ inactivation and thymine dimerization in mutants of bacteriophage T4, *Photochem. Photobiol.*, **4**:555–561, 1965.

Hausmann, R., and Gold, M., The enzymatic methylation of ribonucleic acid and deoxyribonucleic acid, 9. Deoxyribonucleic acid methylase in bacteriophage-infected *Escherichia coli, J. Biol. Chem.*, **241**:1985–1994, 1966.

Hausmann, R., and Gomez, B., Amber mutants of bacteriophages T3 and T7 defective in phage-directed deoxyribonucleic acid synthesis, *J. Virology*, **1**:779–792, 1967.

Hawthorne, D. C., and Friis, J., Osmotic-remedial mutants. A new classification for nutritional mutants in yeast, *Genetics*, **50**:829–839, 1964.

Hayashi, M., Hayashi, M. N., and Spiegelman, S., DNA circularity and the mechanism of strand selection in the generation of genetic messages, *Proc. Natl. Acad. Sci. U.S.*, **51**:351–359, 1964.

Helinski, D. R., and Yanofsky, C., A genetic and biochemical analysis of second site reversion, *J. Biol. Chem.*, **238**:1043–1048, 1963.

Helling, R. B., Selection of a mutant of *Escherichia coli* which has high mutation rates, *J. Bacteriol.*, **96**:975—980, 1968.

Hershey, A. D., Mutation of bacteriophages with respect to type of plaque, *Genetics*, **31**:620—640, 1946.

Hershey, A. D., and Chase, M., Independent functions of viral protein and nucleic acid in growth of bacteriophage, *J. Gen. Physiol.*, **36**:39—56, 1952.

Hershey, A. D., and Rotman, R., Genetic recombination between host range and plaque-type mutants of bacteriophage in single bacterial cells, *Genetics*, **34**:44—71, 1949.

Hertel, R., The occurrence of three allelic markers in one particle of phage T4, *Z. Vererbungslehre*, **94**:436—441, 1963.

Hertel, R., Gene function of heterozygotes in phage T4, *Z. Vererbungslehre*, **96**:105—115, 1965.

Hessler, A. Y., Acridine-resistant mutants of T2H bacteriophage, *Genetics*, **48**:1107—1119, 1963.

Hessler, A. Y., Baylor, M. B., and Baird, J. P., Acridine sensitivity of bacteriophage T2H in *Escherichia coli*, *J. Virology*, **1**:543—549, 1967.

Hill, C. W., Foulds, J., Soll, L., and Berg, P., Instability of a missense suppressor resulting from a duplication of genetic material, *J. Mol. Biol.*, **39**:563—581, 1969.

Hill, R. F., The stability of spontaneous and ultraviolet-induced reversions from auxotrophy in *Escherichia coli*, *J. Gen. Microbiol.*, **30**:289—297, 1963.

Hill, R. F., Ultraviolet-induced lethality and reversion to prototrophy in *Escherichia coli* strains with normal and reduced dark repair ability, *Photochem. Photobiol.*, **4**:563—568, 1965.

Hirota, Y., The effect of acridine dyes on mating type factors in *Escherichia coli*, *Proc. Natl. Acad. Sci. U.S.*, **46**:57—64, 1960.

Holliday, R., A mechanism for gene conversion in fungi, *Genetical Res.*, **5**:282—304, 1964.

Holmes, A. J., and Eisenstark, A., The mutagenic effect of thymine-starvation on *Salmonella typhimurium*, *Mutation Res.*, **5**:15—21, 1968.

Horiuchi, K., Lodish, H. F., and Zinder, N. D., Mutants of the bacteriophage f2, 6. Homology of temperature-sensitive and host-dependent mutants, *Virology*, **28**:438—477, 1966.

Horn, E. E., and Herriott, R. M., The mutagenic action of nitrous acid on "single-stranded" (denatured) *Hemophilus* transforming DNA, *Proc. Natl. Acad. Sci. U.S.*, **48**:1409—1416, 1964.

Horowitz, N. H. and MacLeod, H., The DNA content of Neurospora nuclei, *Microb. Genetics Bull.*, **17**:6—7, 1960.

Howard, B. D., and Tessman, I., Identification of the altered bases in mutated single-stranded DNA, 2. *In vivo* mutagenesis by 5-bromodeoxyuridine and 2-aminopurine, *J. Mol. Biol.*, **9**:364—371, 1964*a*.

Howard, B. D., and Tessman, I., Identification of the altered bases in mutated single-stranded DNA, 3. Mutagenesis by ultraviolet light, *J. Mol. Biol.*, **9**:372—375, 1964*b*.

Howard-Flanders, P., and Boyce, R. P., DNA repair and genetic recombination: studies on mutants of *Escherichia coli* defective in these processes, *Radiation Res.*, Suppl., **6**:156—184, 1966.

Howard-Flanders, P., Boyce, R. P., and Theriot, L., Three loci in *Escherichia coli* K-12 that control the excision of pyrimidine dimers and certain other mutagen products from DNA, *Genetics*, **53**:1119—1136, 1966.

Howarth, S., Resistance to the bactericidal effect of ultraviolet radiation conferred on enterobacteria by the colicine factor *colI*, *J. Gen. Microbiol.*, **40**:43—55, 1965.

Howarth, S., Increase in frequency of ultraviolet-induced mutation brought about by the colicine factor *colI* in *Salmonella typhimurium, Mutation Res.,* 3:129–134, 1966.

Imamoto, F., Ito, J., and Yanofsky, C., Polarity in the tryptophan operon of *E. coli, Cold Spring Harbor Symp. Quant. Biol.,* 31:235–249, 1966.

Imamoto, F., and Yanofsky, C., Transcription of the tryptophan operon in polarity mutants of *Escherichia coli,* 1. Characterization of the tryptophan messenger RNA of polar mutants, *J. Mol. Biol.,* 28:1–23, 1967*a*.

Imamoto, F., and Yanofsky, C., Transcription of the tryptophan operon in polarity mutants of *Escherichia coli,* 2. Evidence for normal production of tryp-mRNA molecules and for premature termination of transcription, *J. Mol. Biol.,* 28:25–35, 1967*b*.

Inge-Vechtomov, S. G., and Pavlenko, V. V., Triallelic complementation and the subunit structure of enzymes, *Nature,* 222:1078–1079, 1969.

Inman, R. B., Denaturation maps of the left and right sides of the lambda DNA molecule determined by electron microscopy, *J. Mol. Biol.,* 28:103–116, 1967.

Inouye, M., Akaboshi, E., Tsugita, A., Streisinger, G., and Okada, Y., A frame-shift mutation resulting in the deletion of two base pairs in the lysozyme gene of bacteriophage T4, *J. Mol. Biol.,* 30:39–47, 1967.

Inselburg, J., Formation of deletion mutations in recombination-deficient mutants of *Escherichia coli, J. Bacteriol.,* 94:1266–1267, 1967.

Iseki, S., and Sakai, T., Artificial transformation of O antigens in Salmonella E group, 1. Transformation by antiserum and bacterial autolysate, *Proc. Japan Acad.,* 29:121–126, 1953.

Itikawa, H., Baumberg, S., and Vogel, H. J., Enzymic basis for a genetic suppression: *N*-acetylglutamic γ-semialdehyde in enterobacterial mutants, *Biochim. Biophys. Acta,* 159:547–550, 1968.

Itikawa, H., and Demerec, M., Ditto deletions in the *cysC* region of the *Salmonella* chromosome, *Genetics,* 55:63–68, 1967.

Iyer, V. N., and Szybalski, W., The mechanism of chemical mutagenesis, 1. Kinetic studies on the action of triethylene melamine (TEM) and azaserine, *Proc. Natl. Acad. Sci. U.S.,* 44:446–456, 1958.

Jacob, F., Mutation d'un bacteriophage induite par l'irradiation des seules bactéries-hôtes avant l'infection, *C. R. Acad. Sci.,* 238:732–734, 1954.

Jacob. F., and Monod, J., Genetic regulatory mechanisms in the synthesis of proteins, *J. Mol. Biol.,* 3:318–356, 1961.

Jacob, F., Ullmann, A., and Monod, J., Le promoteur, element genetique necessaire a l'expression d'un operon, *C. R. Acad. Sci.,* 258:3125–3128, 1964.

Jacob, F., and Wollman, E. L., *Sexuality and the Genetics of Bacteria,* Academic Press, New York, 1961.

Jagger, J., Wise, W. C., and Stafford, R. S., Delay in growth and division induced by near ultraviolet radiation in *Escherichia coli* B and its role in photoprotection and liquid holding recovery, *Photochem. Photobiol.,* 3:11–24, 1964.

Janion, C., and Shugar, D., Mutagenicity of hydroxylamine: reaction with analogues of cytosine, 5(6)-substituted cytosines and some 2-keto-4-ethoxypyrimidines, *Acta Biochim. Polon.,* 12:337–355, 1965*a*.

Janion, C., and Shugar, D., Reaction of hydroxylamine with 5-substituted cytosines, *Biochem. Biophys. Res. Commun.,* 18:617–622, 1965*b*.

Jehle, H., Replication of double-stranded nucleic acids, *Proc. Natl. Acad. Sci. U.S.,* 53:1451–1455, 1965.

Jehle, H., Ingerman, M. L., Shirven, R. M., Parke, W. C., and Salyers, A. A., Replication of nucleic acids, *Proc. Natl. Acad. Sci. U.S.,* 50:738–746, 1963.

Jennings, H. S., Formation, inheritance and variation of the teeth in *Difflugia corona*, a study of the morphogenic activities of rhizopod protoplasm, *J. Exp. Zool.*, **77**:287–336, 1937.

Jensen, R. A., and Haas, F. L., Analysis of ultraviolet light-induced mutagenesis by DNA transformation in *Bacillus subtilis, Proc. Nat. Acad. Sci. U.S.*, **50**:1109–1116, 1963.

Jinks, J. L., Internal suppressors of the *h*III and *tu45* mutants of bacteriophage T4, *Heredity*, **16**:241–254, 1961.

Johns, H. E., LeBlanc, J. C., and Freeman, K. B., Reversal and deamination rates of the main ultraviolet photoproduct of cytidylic acid, *J. Mol. Biol.*, **13**:849–861, 1965.

Johnson, H. G., and Bach, M. K., Apparent suppression of mutation rates in bacteria by spermine, *Nature*, **208**:408–409, 1965.

Johnson, H. G., and Bach, M. K., The antimutagenic action of polyamines: suppression of the mutagenic action of an *E. coli* mutator gene and of 2-aminopurine, *Proc. Nat. Acad. Sci. U.S.*, **55**:1453–1456, 1966.

Jones, R. T., Brimhall, B., Huisman, T. H. J., Kleihauer, E., and Betke, K., Hemoglobin Freiburg: abnormal hemoglobin due to deletion of a single amino acid, *Science*, **154**:1024–1027, 1966.

Jordan, E., The location of the *b2* delection of bacteriophage λ, *J. Mol. Biol.*, **10**:341–344, 1964.

Jordan, E., and Saedler, H., Polarity of amber mutations and suppressed amber mutations in the galactose operon of *E. coli, Molec. Gen. Genetics*, **100**: 283–295, 1967.

Jordan, E., Saedler, H., and Starlinger, P., Strong polar mutations in the transferase gene of the galactose operon in *E. coli, Molec. Gen. Genetics*, **100**: 296–306, 1967.

Jordan, E., and Saedler, H., and Starlinger, P., 0° and strong-polar mutations in the *gal* operon are insertions, *Molec. Gen. Genetics*, **102**:353–363, 1968.

Josse, J., Kaiser, A. D., and Kornberg, A., Enzymatic synthesis of deoxyribonucleic acid, 8. Frequencies of nearest neighbor base sequences in deoxyribonucleic acid, *J. Biol. Chem.*, **236**: 864–875, 1961.

Jyssum, K., Observations on two types of genetic instability in *Escherichia coli, Acta Path. Microbiol. Scand.*, **48**: 113–120, 1960.

Jyssum, K., Mutator factor in *Neisseria meningitidis* associated with increased sensitivity to ultraviolet light and defective transformation, *J. Bacteriol.*, **96**:165–172, 1968.

Jyssum, K., and Jyssum, S. Isolation of variants with increased mutability from *Neisseria meningitidis, Acta Path. Microb. Scand.*, **74**:93–100, 1968.

Kada, T., Mutator action induced by ultraviolet irradiation and λ-lysogenization in *Escherichia coli* K12, *Proc. XXI Intl. Cong. Genetics*, **1**:75, 1968.

Kada, T., and Marcovich, H., Sur le siège initial de l'action mutagène des rayons X et des ultraviolets chez *Escherichia coli* K12, *Ann. Inst. Pasteur*, **105**:989–1006, 1963.

Kanazir, D., The apparent mutagenicity of thymine deficiency, *Biochim. Biophys. Acta*, **30**: 20–23, 1958.

Kaney, A. R., and Atwood, K. C., Radiomimetic action of polyimine chemisterilants in *Neurospora, Nature*, **201**: 1006–1008, 1964.

Kaplan, R. W., Mutations by photodynamic action in *Bacterium prodigiosum, Nature*, **163**: 573–574, 1949.

Kaplan, R. W., Auslösung von Phagenresistenzmutationen bei *Bakterium coli* durch Erythrosin mit und ohne Belichtung, *Naturwiss.*, **37**: 308, 1950a.

Kaplan, R. W., Photodynamische Auslösung Mutationen in den Sporen von *Penicillium notatum, Planta,* **38**: 1–11, 1950*b*.

Kaplan, R. W., Dose-effect curves of *s*-mutation and killing in *Serratia marscens, Arch. Microbiol.,* **24**: 60–79, 1956.

Kaplan, R. W., Spontane Mutation von einer Monoauxotrophie zu einer Anderen in einem Schritt (Auxotrophiesprungmutation), *Z. Vererbungslehre,* **92**: 21–27, 1961.

Kaplan, R. W., Photoreversion von vier Gruppen UV-induzierter Mutationen zur Giftresistenz im Nichtphotoreaktivierbaren *E. coli, Photochem. Photobiol.,* **2**: 461–470, 1963.

Kaplan, R. W., Mischklone U.V.-induzierter Mutatenten beim Phagen Kappa, *Photochem. Photobiol.,* **5**:261–264, 1966.

Kaplan, R. W., and Kaplan, C., Influence of water content on UV-induced *s*-mutation and killing in *Serratia, Exp. Cell Res.,* **11**:378–392, 1956.

Kaplan, R. W., Winkler, U., and Wolf-Ellmauer, H., Induction and reversion of *c*-mutations by irradiation of the extracellular x-phage of *Serratia, Nature,* **186**: 330–331, 1960.

Kaplan, R. W., and Witt, G., Photoreversion und Photoprotektion bei UV-induzierten Mutationen des nichtphotoreaktivierbaren Bakteriums *E. coli* phr⁻, *Z. Vererbungslehre*, **97**:209–217, 1965.

Kaplan S., The genetic code, *Sci. Prog. Oxf.,* **55**:223–238, 1967.

Kaplan, S., Stretton, A. O. W., and Brenner, S., *Amber* suppressors: efficiency of chain propagation and suppressor specific amino acids, *J. Mol. Biol.,* **14**:528–533, 1966.

Karam, J. D., and Speyer, J. F., Mutagenic DNA polymerase, *Fed. Proc.,* **25**:708, 1966.

Katritzky, A. R., and Waring, A. J., Tautomeric azines, Part I: The tautomerization of 1-methyluracil and 5-bromo-1-methyluracil, *J. Chem. Soc.,* 1540–1544, 1962.

Kellenberger, E., and Weigle, J., Etude au moyen des rayons ultraviolet de l'interaction entre bacteriophage tempere et bacterie hote, *Biochim. Biophys. Acta,* **30**:112-124, 1958.

Kellenberger, G., Arber, W., and Kellenberger, E., Eigenschaften UV-bestrahlter λ Phagen, *Z. Naturforsch.*, **14B**:615–629, 1959.

Kellenberger, G., Zichichi, M. L., and Epstein, H. T., Heterozygosis and recombination of bacteriophage λ, *Virology,* **17**:44–55, 1962.

Kellenberger, G., Zichichi, M. L., and Weigle, J., A mutation affecting the DNA content of bacteriophage lambda and its lysogenizing properties, *J. Mol. Biol.,* **3**:399–408, 1961.

Kellogg, D. A., Doctor, B. P., Loebel, J. E., and Nirenberg, M. W., RNA codons and protein synthesis, 9. Synonym codon recognition by multiple species of valine-, alanine-, and methionine-sRNA, *Proc. Natl. Acad. Sci. U.S.,* **55**: 912–919, 1966.

Kelner, A., and Halle, S., Mutagenesis by visible light in mutable strain of *Escherichia coli, Bact. Proc.,* 67, 1960.

Kennell, D., Titration of the gene sites on DNA by DNA-RNA hybridization, 2. The *Escherichia coli* chromosome, *J. Mol. Biol.,* **34**:85–103, 1968.

Kilbey, B. J., Specificity in the photoreactivation of premutational damage induced in *Neurospora crassa* by ultraviolet, *Molec. Gen. Genetics,* **100**:159–165, 1967.

Kilbey, B. J., and de Serres, F. J., Quantitative and qualitative aspects of photoreactivation of premutational ultraviolet damage at the *ad-3* loci of *Neurospora crassa, Mutation Res.,* **4**:21–29, 1967.

Kimura, M., On the evolutionary adjustment of spontaneous mutation rates, *Genetical Res.*, **9**:23–34. 1967.

Kimura, M., Maruyama, T., and Crow, J. F., The mutation load in small populations, *Genetics*, **48**:1303–1312, 1963.

Kirchner, C. E. J., The effects of the mutator gene on molecular changes and mutation in *Salmonella typhimurium*, *J. Mol. Biol.*, **2**:331–338, 1960.

Kirchner, C. E. J., and Rudden, M. J., Location of a mutator gene in *Salmonella typhimurium* by cotransduction, *J. Bacteriol.*, **92**:1453–1456, 1966.

Kleinwächter, V., and Koudelka, J., Thermal denaturation of deoxyribonucleic acid–acridine orange complexes, *Biochim. Biophys. Acta*, **91**:539–540, 1964.

Kneser, H., Absence of K-reactivation in a recombination-deficient mutant of *E. coli*, *Biochem. Biophys. Res. Commun.*, **22**:383–387, 1966.

Kneser, H., Metzger, K., and Sauerbier, W., Evidence of different mechanisms for ultraviolet reactivation and "ordinary host cell reactivation" of phage λ, *Virology*, **27**:213–221, 1965.

Koch, A. L., and Miller, C., A mechanism for keeping mutations in check, *J. Theoret., Biol.*, **8**:71–80, 1965.

Koch, J. P., The increased mutation frequency of a gene of *E. coli* upon being rendered constitutive, *Fed. Proc.*, **27**:700, 1968.

Kohiyama, M., and Ikeda, Y., Appearance of new deficient characters in streptomycin-resistant mutants of *Bacillus subtilis, Nature*, **187**:168–169, 1960.

Kondo, S., and Jagger, J., Action spectra for photoreactivation of mutation to prototrophy in strains of *Escherichia coli* possessing and lacking photo-reactivating-enzyme activity, *Photochem. Photobiol.*, **5**:189–200, 1966.

Kondo, S., and Kato, T., Action spectra for photoreaction of killing and mutation to prototrophy in U.V.–sensitive strains of *Escherichia coli* possessing and lacking photoreactivating enzyme, *Photochem. Photobiol.*, **5**:827–837, 1966.

Kornberg, A., The active center of DNA polymerase, *Science*, **163**:1410–1418, 1969.

Kotaka, T., and Baldwin, R. L., Effects of nitrous acid on the dAT copolymer as a template for DNA polymerase, *J. Mol. Biol.*, **9**:323–339, 1964.

Kozinski, A. W., Kozinski, P. B., and Shannon, P., Replicative fragmentation in T4 phage: Inhibition by chloramphenicol, *Proc. Natl. Acad. Sci. U.S.*, **50**:746–753, 1963.

Kozinski, A. W., and Lin, T. H., Early intracellular events in the replication of T4 phage DNA, 1. Complex formation of replicative DNA, *Proc. Natl. Acad. Sci. U.S.*, **54**:273–278, 1965.

Kramer, G., Wittmann, H. G., and Schuster, H., Die Eizeugung von Mutaten des Tabakmosaikvirus durch den Einbau von Fluoruracil in die Virusnucleinsäure, *Z. Naturforsch.*, **19**B:46–51, 1964.

Krauch, C. H., Drämer, D. M., and Wacker, A., Zum Wirkungsmechanismus photodynamischer Fucomarine. Photoreaktion von Psoralen–(4–^{14}C) mit DNS, RNS, Homopolynucleotiden und Nucleosiden, *Photochem. Photobiol.*, **6**:341–354, 1967.

Krieg, D. R., Induced reversion of T4rII mutants by ultraviolet irradiation of extracellular phage, *Virology*, **9**:215-227, 1959.

Krieg, D. R., Specificity of chemical mutagenesis. *Prog. Nucleic Acid Res.*, **2**:125–168, 1963a.

Kreig, D. R., Ethyl methylsulfonate-induced reversion of bacteriophage T4rII mutants, *Genetics*, **48**:561–580, 1963b.

Kubitschek, H. E., The error hypothesis of mutation, *Science,* 131:730, 1960,

Kubitschek, H. E., Mutation without segregation, *Proc. Natl. Acad. Sci. U.S.,* 52:1374–1381, 1964.

Kubitschek, H. E., Mutation without segregation in bacteria with reduced dark repair ability, *Proc. Natl. Acad. Sci. U.S.,* 55:269–277, 1966.

Kubitschek, H. E., Mutagenesis by near-visible light, *Science,* 155:1545–1546, 1967.

Kubitschek, H. E., and Bendigkeit, H. E., Mutation in continuous cultures, 1. Dependence of mutational response upon growth-limiting factors, *Mutation Res.,* 1:113–120, 1964*a*.

Kubitschek, H. E., and Bendigkeit, H. E., Mutation in continuous cultures, 2. Mutations induced with ultraviolet light and 2-aminopurine, *Mutation Res.,* 1:209–218, 1964*b*.

Kubitschek, H. E., and Gustafson, L. A., Mutation in continuous cultures, 3. Mutational responses in *Escherichia coli, J. Bacteriol.,* 88:1595–1597, 1964.

Kubitschek, H. E., and Henderson, T. R., DNA replication, *Proc. Natl. Acad. Sci. U.S.,* 55:512–519, 1966.

Kuo, T. -T., and Stocker, B. A. D., Suppression of proline requirement of *proA* and *proAB* deletion mutants in *Salmonella typhimurium* by mutation to arginine requirement, *J. Bacteriol.,* 98:593–598, 1969.

Kuwano, M., Endo, H., and Ohnishi, Y., Mutations to spectinomycin resistance which alleviate the restriction of an amber suppressor by streptomycin resistance, *J. Bacteriol.,* 97:940–943, 1969.

Laird, C. D., and Bodmer, W. F., 5-Bromouracil utilization by *Bacillus subtilis, J. Bacteriol,* 94:1277–1278, 1967.

Lamola, A. A., and Yamane, T., Sensitized photodimerization of thymine in DNA, *Proc. Natl. Acad. Sci. U.S.,* 58:443–446, 1967.

Langridge, J., and Campbell, J. H., Classification and intragenic position of mutations in the β-galactosidase gene of *Escherichia coli, Molec. Gen. Genetics,* 103: 339–347, 1969.

Latarjet, R., Mutation induite chez un virus par irradiation ultraviolette de cellules infectées, *C. R. Acad. Sci.,* 228:1354–1357, 1949.

Latarjet, R., Mutations induites chez un bactériophage, *Acta Unio. Intern. contra Cancrum.,* 10:136, 1954.

Lawley, P. D., Effects of some chemical mutagens and carcinogens on nucleic acids, *Prog. Nucleic Acid Res. Mol. Biol.,* 5:89–131, 1966.

Lawley, P. D., Reaction of hydroxylamine at high concentration with deoxycytidine or with polycitidylic acid: evidence that substitution of amino groups in cytosine residues by hydroxylamine is a primary reaction, and the possible relevence to hydroxylamine mutagenesis, *J. Mol. Biol.,* 24:75–81, 1967.

Lawley, P. D., Methylation of DNA by N-methyl-N-nitrosourethane and N-methyl-N-nitroso-N′-nitroguanidine, *Nature,* 218:580–581, 1968.

Lawley, P. D., and Brookes, P., Acidic dissociation of 7:9-dialkylguanines and its possible relation to mutagenic properties of alkylating agents, *Nature,* 192:1081–1082, 1961.

Lawley, P. D., and Brookes, P., Ionization of DNA bases or base analogues as a possible explanation of mutagenesis, with special reference to 5-bromodeoxy-uridine, *J. Mol. Biol.,* 4:216–219, 1962.

Lawley, P. D., and Brookes, P., Further studies on the alkylation of nucleic acids and their constituent nucleotides, *Biochem. J.,* 89:127–138, 1963.

Lea, D. E., and Coulson, C. A., The distribution of numbers of mutants in bacterial populations, *J. Genetics,* 49:264–285, 1949.

Lederberg, E. M., Cavalli-Sforza, L., and Lederberg, J., Interaction of strepto-mycin and a suppressor for galactose fermentation in *E. coli* K-12, *Proc. Natl. Acad. Sci. U.S.*, **51**:678–682, 1964.

Leonard, N. J., and Laursen, R. A., Synthesis and properties of analogs of adenosine diphosphate, adenosine triphosphate, and nicotinamide-adenine dinucleotide derived from 3-β-D-ribofuranosyladenine (3-isoadenosine), *Biochem.*, **4**:365–376, 1966.

Lerman, L. S., Structural considerations in the interaction of DNA and acridines, *J. Mol. Biol.*, **3**:18–30, 1961.

Lerman, L. S., The structure of the DNA-acridine complex, *Proc. Natl. Acad. Sci. U.S.*, **49**:94–102, 1963.

Lerman, L., Amino group reactivity in DNA-acridine complexes, *J. Mol. Biol.*, **10**:367–380, 1964*a*.

Lerman, L. S., Acridine mutagens and DNA structure, *J. Cell. Comp. Physiol.*, **64**:Suppl. 1:1–18, 1964*b*.

Levins, R., Theory of fitness in a heterogeneous environment, 6. The adaptive significance of mutation, *Genetics*, **56**:163–178, 1967.

Levisohn, R., Genotypic reversion by hydroxylamine: The orientation of guanine-hydroxymethylcytosine at mutated sites in phage T4rII, *Genetics*, **55**:345–348, 1967.

Lewis, E. B., Pseudoallelism and gene evolution, *Cold Spring Harbor Symp. Quant. Biol.*, **16**:159–172, 1951.

Lieb, M., Forward and reverse mutation in a histidine-requiring strain of *Escherichia coli*, *Genetics*, **36**:460–477, 1951.

Lieb, M., Enhancement of ultraviolet-induced mutation in bacteria by caffeine, *Z. Vererbungslehre*, **92**:416–429, 1961.

Likover, T. E., and Kurland, C. G., The contribution of DNA to translation errors induced by streptomycin *in vitro*, *Proc. Natl. Acad. Sci. U.S.*, **58**:2385–2392, 1967.

Litman, R. M., and Pardee, A. B., Production of bacteriophage mutants by a dis-turbance of deoxyribonucleic acid metabolism, *Nature*, **178**:529–531, 1956.

Litman, R. M., and Pardee, A. B., The induction of mutants of bacteriophage T2 by 5-bromouracil, 4. Kinetics of bromouracil-induced mutagenesis, *Biochim. Biophys. Acta*, **42**:131-140, 1960.

Loprieno, N., Abbondandola, A., Bonatti, S., and Guglielminetti, R., Analysis of the genetic instability induced by nitrous acid in *Schizosaccharomyces pombe*, *Genetical Res.*, **12**:45–54, 1968.

Loprieno, N., Bonatti, S., Abbondandolo, A., and Guglielminetti, R., The nature of spontaneous mutations during vegetative growth in *Schizosaccharomyces pombe*, *Molec. Gen. Genetics*, **104**:40–50, 1969.

Lorena, M. J., Inouye, M., and Tsugita, A., Studies on the lysozyme from the bacteriophage T4 *eJD7eJD4*, carrying two frameshift muations, *Molec. Gen. Genetics*, **102**:69–78, 1968.

Louarn, J. M., and Sicard, A. M., Transmission of genetic information during transformation in *Diplococcus pneumoniae*, *Biochem. Biophys. Res. Commun.*, **30**:683–689, 1968.

Loveless, A., Increased rate of plaque-type and host-range mutation following treatment of bacteriophage *in vitro* with ethyl methane sulphonate, *Nature*, **181**:1212–1213, 1958.

Loveless, A., *Genetic and Allied Effects of Alkylating Agents*, Pennsylvania State University Press, University Park, 1966.

Loveless, A., and Howarth, S., Mutation of bacteria at high levels of survival by ethyl methane sulphonate, *Nature*, **184**:1780–1782, 1959.

Loveless, A., Possible relevance of 0–6 alkylation of deoxyguanosine to the mutagenicity and carcinogenicity of nitrososamines and nitrosamides, *Nature*, 223:206–207, 1969.

Ludlum, D. B., and Wilhelm, R. C., Ribonucleic acid polymerase reactions with methylated polycytidylic acid templates, *J. Biol. Chem.*, 243:2750–2753, 1968.

Luria, S. E., The frequency distribution of spontaneous bacteriophage mutants as evidence for the exponential rate of phage reproduction, *Cold Spring Harbor Symp. Quant. Biol.*, 16:463–470, 1951.

Luria, S. E., and Darnell, J. E., *General Virology*, John Wiley and Sons, New York, 1967.

Luria, S. E., and Delbrück, M., Mutations of bacteria from virus sensitivity to virus resistance, *Genetics*, 28:491–511, 1943.

Luzzati, D., The action of nitrous acid on transforming desoxyribonucleic acids, *Biochem. Biophys. Res. Commun.*, 9:508–516, 1962.

Luzzati, V., Masson F., and Lerman, L. S., Interaction of DNA and proflavine: a small-angle X-ray scattering study, *J. Mol. Biol.*, 3:634–639, 1961.

MacHattie, L. A., Ritchie, D. A., Thomas, C. A., and Richardson, C. C., Terminal repetition in permuted T2 bacteriophage DNA molecules, *J. Mol. Biol.*, 23:355–363, 1967.

Maestre, M. F., and Tinoco, I., Optical rotary dispersion of viruses, *J. Mol. Biol.*, 23:323–335, 1967.

Magni, G. E., The origin of spontaneous mutations during meiosis, *Proc. Natl. Acad. Sci. U.S.*, 50:975–980, 1963.

Magni, G. E., and Puglisi, P. P., Mutagenesis of super-suppressors in yeast, *Cold Spring Harbor Symp. Quant. Biol.*, 31:699–704, 1966.

Magni, G. E., von Borstel, R. C., and Sora, S., Mutagenic action during meiosis and antimutagenic action during mitosis by 5-aminoacridine in yeast, *Mutation Res.*, 1:227–230, 1964.

Magni, G. E., von Borstel, R. C., and Steinberg, C. M., Super-suppressors as addition-deletion mutations, *J. Mol. Biol.*, 16:568–570, 1966.

Mahler, H. R., and Baylor, M. B., Effects of steroidal diamines on DNA duplication and mutagenesis, *Proc. Natl. Acad. Sci. U.S.*, 58:256–263, 1967.

Malamy, M. H., Frameshift mutations in the lactose operon of *E. coli*, *Cold Spring Harbor Symp. Quant. Biol.*, 31:189–201, 1966.

Malling, H. V., The mutagenicity of the acridine mustard (ICR-170) and the structurally related compounds in *Neurospora*, *Mutation Res.*, 4:265–274, 1967.

Malling, H. V., and de Serres, F. J., Correlation between base-pair transition and complementation pattern in nitrous acid-induced *ad-3B* mutants of *Neurospora crassa*, *Mutation Res.*, 5:359–371, 1968*a*.

Malling, H. V., and de Serres, F. J., Identification of genetic alterations induced by ethyl methanesulfonate in *Neurospora crassa*, *Mutation Res.*, 6:181–193, 1968*b*.

Mandell, J. D., and Greenberg, J., A new chemical mutagen for bacteria, 1-methyl-3-nitro-1-nitrosoguanidine, *Biochem. Biophys. Res. Commun.*, 3:575–577, 1961.

Margolies, M. N., and Goldberger, R. F., Correlation between mutation type and the production of cross-reacting material in mutants of the *A* gene of the histidine operon in *Salmonella typhimurium*, *J. Bacteriol.*, 95:507–519, 1968.

Margolin, P., and Mukai, F. H., The pattern of mutagen-induced back mutations in *Salmonella typhimurium*, *Z. Vererbungslehre*, 92:330–335, 1961.

242 BIBLIOGRAPHY

Margolin, P., and Mukai, F. H., A model for mRNA transcription suggested by some characteristics of 2-aminopurine mutagenesis in *Salmonella, Proc. Natl. Acad. Sci. U.S.,* **55**:282–289, 1966.

Marmur, J., and Greenspan, C. M., Transcription *in vivo* of DNA from bacteriophage SP8, *Science,* **142**:387–389, 1963.

Marmur, J., and Grossman, L., Ultraviolet light induced linking of deoxyribonucleic acid strands and its reversal by photoreactivating enzyme, *Proc. Natl. Acad. Sci. U.S.,* **47**:778–787, 1961.

Martin, R. G., Frameshift mutants in the histidine operon of *Salmonella typhimurium, J. Mol. Biol.,* **26**:311–328, 1967.

Martin, R. G., Polarity in relaxed strains of *Salmonella typhimurium, J. Mol. Biol.,* **31**:127–134, 1968.

Martin, R. G., Silbert, D. F., Smith, D. W. E., and Whitfield, H. J., Polarity in the histidine operon, *J. Mol. Biol.,* **21**:357–369, 1966*a.*

Martin, R. G., Whitfield, H., Berkowitz, D., and Voll, M., A molecular model of the phenomenon of polarity, *Cold Spring Harbor Symp. Quant. Biol.,* **31**:215–220, 1966*b.*

Mathews, M. M., Comparative study of lethal photosensitization of *Sarcina lutea* by 8-methoxypsoralen and by toluidine blue, *J. Bacteriol.,* **85**:322–328, 1963.

Matsuura, T., and Saito, I., Photosensitized oxidation of hydroxylated purines, *Chem. Commun.,* 693–694, 1967.

Mauss, Y., Chambron, J., Daune, M., and Benoit, H., Etude morphologique par diffusion de la lumiere du complexe formé par le DNA et la proflavine, *J. Mol. Biol.,* **27**:579–589, 1967.

McCalla, D. R., Reaction of N-methyl-N'-nitro-N-nitroguanidine and N-methyl-N-nitrose-*p*-toluenesulfonamide with DNA *in vitro, Biochim. Biophys. Acta,* **155**:114–120, 1968.

McClain, W. H., and Champe, S. P., Detection of a peptide determined by the *r*II B cistron of phage T4, *Proc. Natl. Acad. Sci. U.S.,* **58**:1182–1188, 1967.

McClintock, B., The control of gene action in maize, *Brookhaven Symp. Biol.,* **18**:162–184, 1965.

McFall, E., and Stent, G. S., Three star mutants of coliphage T2, *J. Gen Microbiol.,* **18**:346–369, 1958.

McGavin, S., Interallelic complementation and allostery, *J. Mol. Biol.,* **37**:239–242, 1968.

Melnick, J. L., Crowther, D., and Barrera-Oro, J., Rapid development of drug-resistant mutants of poliovirus, *Science,* **134**:557, 1961.

Meselson, M., On the mechanism of genetic recombination between DNA molecules, *J. Mol. Biol.,* **9**:734–745, 1964.

Meselson, M., The molecular basis of genetic recombination in *Heritage from Mendel,* edited by R. A. Brink, pp. 81–103, University of Wisconsin Press, Madison, 1967.

Meselson, M., and Weigle, J. J., Chromosome breakage accompanying genetic recombination in bacteriophage, *Proc. Natl. Acad. Sci. U.S.,* **47**:857–868, 1961.

Meselson, M., and Yuan, R., DNA restriction enzyme from *E. coli, Nature,* **217**:1110–1114, 1968.

Michels, C. A., and Zipser, D., Mapping of polypeptide reinitiation sites within the β-galactosidase structural gene, *J. Mol. Biol.,* **41**:341–347, 1969.

Michelson, A. M., and Grunberg-Manago, M., Polynucleotide analogues, 5. "Nonsense" bases, *Biochim. Biophys. Acta,* **91**:92–104, 1964.

Michelson, A. M., and Monny, C., Polynucleotide analogues, 9. Polyxanthylic acid, *Biochim. Biophys. Acta,* **129**:460–474, 1966.

Michelson, A. M., Monny, C., Laursen, R. A., and Leonard, N. J., Polynucleotide analogues, 8. Poly 3-isoadenylic acid, *Biochim. Biophys. Acta*, **119**:258—267, 1966.

Miura, A., and Tomizawa, J., Studies on radiation-sensitive mutants of *E. coli*, 3. Participation of the rec system in induction of mutations by ultraviolet irradiation, *Molec. Gen. Genetics*, **103**:1—10, 1968.

Miyake, T., Mutator factor in *Salmonella typhimurium, Genetics*, **45**:11—14, 1960.

Mohn, G., Korrelation zwischen verminderter Reparaturfähigkeit für UV-Läsionen und hoher Spontanmutabilitat eines Mutatorstammes von *E. coli* K-12, *Molec. Gen. Genetics*, **101**:43—50, 1968.

Mohn, G., and Kaplan, R. W., Beeinflussung der spontanen und chemisch induzierten Mutabilität bei *E. coli* durch Änderung des genetischen Hintergrunds, *Molec. Gen. Genetics*, **99**:191—202, 1967.

Morikawa, N., and Imamoto, F., on the degradation of messenger RNA for the tryptophan operon in *Escherichia coli, Nature*, **223**:37—40, 1969.

Morse, D. E., Mosteller, R., Baker, R. F., and Yanofsky, C., Direction of *in vivo* degradation of tryptophan messenger RNA — a correction, *Nature*, **223**:40—43, 1969.

Mosig, G., Genetic recombination in bacteriophage T4 during replication of DNA fragments, *Cold Spring Harbor Symp. Quant. Biol.*, **28**:35—41, 1963.

Mosig, G., Distance separating genetic markers in T4 DNA, *Proc. Natl. Acad. Sci. U.S.*, **56**:1177—1183, 1966.

Mukai, F. H., and Margolin, P., Analysis of unlinked suppressors of an 0° mutation in *Salmonella, Proc. Natl. Acad. Sci. U.S.*, **50**:140—148, 1963.

Mukai, T., The genetic structure of natural populations of *Drosophila melanogaster*, 1. Spontaneous mutation rate of polygenes controlling viability, *Genetics*, **50**:1—19, 1964.

Muller, H. J., An *Oenothera*-like case in *Drosophila, Proc. Natl. Acad. Sci. U.S.*, **3**:619—626, 1917.

Muller, H. J. Artificial transmutation of the gene, *Science*, **66**:84—87, 1927.

Muller, H. J., Our load of mutations, *Am. J. Human Genetics*, **2**:111—176, 1950.

Muller, H. J., Advances in radiation mutagenesis through studies on *Drosophila, Progress in Nuclear Energy*, Series VI, Vol. 2, pp. 146—160, Pergamon Press, London, 1959.

Muller, H. J., Carlson, E., and Schalet, A., Mutation by alteration of the already existing gene, *Genetics*, **46**:213—226, 1961.

Musajo, L., Bordın, F., and Bevilacqua, R., Photoreactions at 3655 A linking the 3-4 double bond of fucocoumarins with pyrimidine bases, *Photochem. Photobiol.*, **6**:927—931, 1967.

Musajo, L., Bordin, F., Caporale, G., Marciani, S., and Rigatti, G., Photoreactions at 3655 Å between pyrimidine bases and skin-photosensitizing fucomarins, *Photochem. Photobiol.*, **6**:711-719, 1967.

Nakai, S., and Saeki, T., Induction of mutation by photodynamic action in *Escherichia coli, Genetical Res.*, **5**:158—161, 1964.

Nakata, A., and Stahl, F. W., Further evidence for polarity mutations in bacteriophage T4, *Genetics*, **55**:585—590, 1967.

Naono, S., and Gros, F., Synthèse par *E. coli* d'une phosphatase modifiée en présence d'un analogue pyrimidique, *C. R. Acad. Sci.*, **250**:3889—3891, 1960.

Nasim, A., Repair-mechanisms and radiation-induced mutations in fission yeast, *Genetics*, **59**:327—333, 1968.

Nasim, A., The induction of replicating instabilities by mutagens in *Schizosaccharomyces pombe, Mutation Res.,* 4:753–763, 1967.

Nasim, A., and Auerbach, C., The origin of complete and mosaic mutants from mutagenic treatment of single cells, *Mutation Res.,* 4:1–14, 1967.

Neidhardt, F., and Earhart, C., Phage-induced appearance of a valyl sRNA synthetase activity in *Escherichia Coli, Cold Spring Harbor Symp. Quant. Biol.,* 31:557–563, 1966.

Newton, A., Effect of nonsense mutations on translation of the lactose operon of *Escherichia coli, Cold Spring Harbor Symp. Quant. Biol.,* 31:181–186, 1966.

Newton, A., Re-initiation of polypeptide synthesis and polarity in the *lac* operon of *Escherichia coli, J. Mol. Biol.,* 41:329–339, 1969.

Newton, W. A., Beckwith, J. R., Zipser, D., and Brenner, S., Nonsense mutations and polarity in the *lac* operon of *Escherichia coli, J. Mol. Biol.,* 14:290–296, 1965.

Nirenberg, M. W., and Matthaei, J. H., The dependence of cell-free protein synthesis in *E. coli* upon naturally occurring or synthetic polyribonucleotides, *Proc. Natl. Acad. Sci. U.S.,* 47:1588–1602, 1961.

Nomura, M., and Benzer, S., The nature of the "deletion" mutants in the *r*II region of phage T4, *J. Mol. Biol.,* 3:684–692, 1961.

Novick, A., Mutagens and antimutagens, *Brookhaven Symp. Biol.,* 8:201–215, 1956.

Novick, A., and Szilard, L., Experiments on spontaneous and chemically induced mutations of bacteria growing in the chemostat, *Cold Spring Harbor Symp. Quant. Biol.,* 16:337–343, 1951.

Novick, A., and Weiner, M., Enzyme induction as an all-or-none phenomenon, *Proc. Natl. Acad. Sci. U.S.,* 43:553–566, 1957.

Oeschger, N. S., and Stahl, R. C., Mutations induced in bacteria by an acridine half-mustard ICR191, a non-alkylating aza-acridine ICR364-OH, and by nitrosoguanidine (NG), *Bact. Proc.,* 59, 1967.

Ohtaka, Y., and Spiegelman, S., Translational control of protein synthesis in a cell-free system directed by a polycistronic viral RNA, *Science,* 142:493–497, 1963.

Okada, Y., Terzaghi, E., Streisinger, G., Emrich, J., Inouye, M., and Tsugita, A., A frame-shift mutation involving the addition of two base pairs in the lysozyme gene of phage T4, *Proc. Natl. Acad. Sci. U.S.,* 56:1692–1698, 1966.

Ono, J., Wilson, R. G., and Grossman, L., Effect of ultraviolet light on the template properties of polycytidylic acid, *J. Mol. Biol.,* 11:600–612, 1965.

Orgel, A., and Brenner, S., Mutagenesis of bacteriophage T4 by acridines, *J. Mol. Biol.,* 3:762–768, 1961.

Orgel, A., and Orgel, L.E., Induction of mutations in bacteriophage T4 with divalent manganese, *J. Mol. Biol.,* 14:453–457, 1965.

Orgel, L. E., The chemical basis of mutation, *Adv. Enzymology,* 27:289–346, 1965.

Orias, E., and Gartner, T. K., Suppression of *amber* and *ochre r*II of bacteriophage T4 by streptomycin, *J. Bacteriol.,* 92:2210–2215, 1966.

Otsuji, N., and Aono, H., Effect of mutation to streptomycin resistance on amber suppressor genes, *J. Bacteriol.,* 96:43–50, 1968.

Papirmeister, B., and Davison, C. L., Elimination of sulfur mustard-induced products from DNA of *Escherichia coli, Biochem. Biophys. Res. Commun.,* 17:608–617, 1964.

Pauling, C., The specificity of thymineless mutagenesis, in *Structural Chemistry and Molecular Biology: A Volume Dedicated to Linus Pauling by his*

Students, Colleagues, and Friends, edited by A. Rich and N. Davidson, W. H. Freeman and Co., San Francisco, 1968.

Peacocke, A. R., and Skerrett, J. N. H., The interaction of aminoacridines with nucleic acids, *Trans. Faraday Soc.,* 52:261–279, 1956.

Pearson, M. L., Ottensmeyer, F. P., and Johns H. E., Properties of an unusual photoproduct of u.v. irradiated thymidylyl-thymidine, Photochem. Photobiol., 4:739–747, 1965.

Pearson, M., Whillans, D. W., LeBlanc, J. C., and Johns, H. E., Dependence on wavelength of photoproduct yields in ultraviolet-irradiated poly U, *J. Mol. Biol.,* 20:245–261, 1966.

Perutz, M. F., Kendrew, J. C., and Watson, H. C., Structure and function of haemoglobin, 2. Some relations between polypeptide chain configuration and amino acid sequence, *J. Mol. Biol.,* 13:669–678, 1965.

Perutz, M. F., and Lehmann, H., Molecular pathology of human haemoglobin, *Nature,* 219:902–909, 1968.

Perutz, M. F., Muirhead, H. Cox, J. M., and Goaman, L. C. G., Three-dimensional fourier synthesis of horse oxyhaemoglobin at 2.8 Å resolution: the atomic model, *Nature.* 219:131–139, 1968.

Pestka, S., Marshall, R., and Nirenberg, M., RNA codewords and protein synthesis, 5. Effect of streptomycin on the formation of ribosome-sRNA complexes, *Proc. Natl. Acad. Sci. U.S.,* 53:639–646, 1965.

Pettijohn, D., and Hanawalt, P., Evidence for repair-replication of ultraviolet damaged DNA in bacteria, *J. Mol. Biol.,* 9:395–410, 1964.

Phillips, J. H., and Brown, D. M., The mutagenic action of hydroxylamine, *Prog. Nucleic Acid Res. Mol. Biol.,* 7:349–368, 1967.

Phillips, J. H., Brown, D. M. Adman, R., and Grossman, L., The effects of hydroxylamine on polynucleotide templates for RNA polymerase, *J. Mol. Biol.,* 12:816–828, 1965.

Phillips, J. H., Brown, D. M., and Grossman, L., The efficiency of induction of mutations by hydroxylamine, *J. Mol. Biol.,* 21:405–419, 1966.

Phillips, J. N., Modification of mutation frequency in microorganisms, 1. Preirradiation and postirradiation treatment, *Genetics,* 46:317–322, 1961.

Piechowski, M. M., and Susman, M., Acridine-resistance in phage T4D, *Genetics,* 56:133–148, 1967.

Pierce, B. L. S., The effect of a bacterial mutator gene upon mutation rates in bacteriophage T4, *Genetics,* 54:657–662, 1966.

Pitha, P. M., Huang, W. M., and Ts'o, P. O. P., Physiochemical basis of the recognition process in nucleic acid interactions, 4. Costacking as the cause of mispairing and intercalation in nucleic acid interactions, *Proc. Natl. Acad. Sci. U.S.,* 61:332–339, 1968.

Plough, H. H., Spontaneous mutability in *Drosophila. Cold Spring Harbor Symp. Quant. Biol.,* 9:127–137, 1941.

Pratt, D., and Stent, G. S., Mutational heterozygotes in bacteriophages, *Proc. Natl. Acad. Sci. U.S.,* 45:1507–1515, 1959.

Prell, H. H., Inaktivierung des Salmonella-Bakteriophagen P22 durch Chloroform, 1. Reaktionskinetik, *Arch. Mikrobiol.,* 36:151–168, 1960*a*.

Prell, H. H., Inaktivierung des Salmonella-Bakteriophagen P22 durch Chloroform, 2. Functionsanalyse des chloroforminduzierten Shadens, *Arch. Mikrobiol.,* 37:399–420, 1960*b*.

Price, C. C., Gaucher, G. M., Koneru, P., Shibakawa, R., Sowa J. R., and Yamaguchi, M., Relative reactivities for monofunctional nitrogen mustard alkylation of nucleic acid components, *Biochim. Biophys. Acta,* 166:327–359, 1968.

Pritchard, N. J., Blake, A., and Peacocke, A. R., Modified intercalation model for the interaction of amino acridines and DNA, *Nature,* 212:1360–1361, 1966.

Prozorov, A. A., and Barabanshchikov, B. I., The rate of spontaneous mutational process in a strain of *Bact. subtilis* showing a disturbed recombination ability, *Dokl. Akad. Nauk; SSSR*, 176: 1422–1424, 1967.

Prozorov, A. A., and Barabanshchikov, B. I., the rate of spontaneous mutagenisis and the specificity of mutations initiating in the strains of *Bacillus subtilis* with a reduced capacity of genetic recombination, *Genetics (USSR),* 4 no. 5:80–87, 1968.

Ptashne, M., Isolation of the λ phage repressor, *Proc. Natl. Acad. Sci. U.S.,* 57:306–313, 1967.

Puglisi, P. P., Antimutagenic activity of actinomycin D and basic fuchsine in *Saccharomyces cerevisiae, Molec. Gen. Genetics,* 103:248–252, 1968.

Puglisi, P. P., Mutagenic and antimutagenic effects of acridinium salts in yeast, *Mutation Res.,* 4:289–294, 1967.

Pullman, B., and Pullman, A., La tautomerie des bases purique et pyrimidiques et la theorie des mutations, *Biochim. Biophys. Acta,* 64:403–405, 1962.

Pullman, B., and Pullman, A., *Quantum Biochemistry,* Interscience, New York, 1963.

Rahn, R. O., Shulman, R. G., and Longworth, J. W., The UV-induced triplet state in DNA, *Proc. Natl. Acad. Sci. U.S.,* 53:893–896, 1965.

Rappaport, I., and Wildman, S. G., A kinetic study of local lesion growth on *Nicotiana glutinosa* resulting from tobacco mosaic virus infection, *Virology,* 4:265–274, 1957.

Regös, J., and Szende, K., Suppression of the heat-sensitive mutants of the coliphage lambda, *Virology,* 33:748–749, 1967.

Reid, P., and Berg, P., T4 bacteriophage mutants suppressible by a missense suppressor which inserts glycine in place of arginine for the codon AGA, *J. Virology,* 2:905–914, 1968.

Revel, H. R., Restriction of nonglucosylated T-even bacteriophage: properties of permissive mutants of *Escherichia coli* B and K12, *Virology,* 31:688–701, 1967.

Rhaese, H. J., Chemical analysis of DNA alterations, 3. Isolation and characterization of adenine oxidation products obtained from oligo- and monodeoxyadenlic acids treated with hydroxyl radicals, *Biochim. Biophys. Acta,* 166:311–326, 1966.

Richardson, C. C., Phosphorylation of nucleic acid by an enzyme from T4 bacteriophage-infected *Escherichia coli, Proc. Natl. Acad. Sci. U.S.,* 54:158–165, 1965.

Richardson, C. C., Influence of glucosylation of deoxyribonucleic acid on hydrolysis by deoxyribonucleases of *Escherichia coli, J. Biol. Chem.,* 241:2084–2092, 1966.

Richardson, C. C., Schildkraut, C. L., and Kornberg, A., Studies on the replication of DNA by DNA polymerase, *Cold Spring Harbor Symp. Quant. Biol.,* 28:9–18, 1963.

Ritchie, D. A., Mutagenesis with light and proflavin in phage T4, *Genetical Res.,* 5:168–169, 1964.

Ritchie, D. A., Mutagenesis with light and proflavin in phage T4, 2. Properties of the mutants, *Genetical Res.,* 6:474–478, 1965.

Riva, S. C., Interactions of methylated acridines with DNA, *Biochem. Biophys. Res. Commun.,* 23:606–611, 1966.

Riyasaty, S., and Atkins, J. F., External suppression of a frameshift mutant in *Salmonella, J. Mol. Biol.,* **34**:541–557, 1968.

Riyasaty, S., and Dawson, G. W. P., The recovery of tryptophan A auxotrophs at high frequency in a strain of *Salmonella typhimurium, Genetical Res.,* **10**:127–134, 1967.

Roberts, J. W., and Gussin, G. N., Polarity in an amber mutant of bacteriophage R17, *J. Mol. Biol.,* **30**:565–570, 1967.

Robinson, W. E., Tessman, I., and Gilham, P. T., Demonstration of sequence difference in the ribonucleic acids of bacteriophage MS2, and a mutant of MS2, *Biochemistry,* **8**:483–488, 1969.

Roller, A., Replication and transfer of the DNA of phage T4, *J. Mol. Biol.,* **9**:260–262, 1964.

Ronen, A., Back mutation of leucine-requiring auxotrophs of *Salmonella typhimurium* induced by diethylsulphate, *J. Gen. Microbiol.,* **37**:49–58, 1964.

Ronen, A., Inactivation of phage T4 by ethylmethane sulfonate, *Biochem. Biophys. Res. Commun.,* **33**:190–196, 1968.

Rosen, B., Characteristics of 5-fluorouracil-induced synthesis of alkaline phosphatase, *J. Mol. Biol.,* **11**:845–850, 1965.

Rosner, J., and Barricelli, N. A., Evidence derived from HNO_2 mutagenesis that only one of the two DNA strands injected by phage T4 transmits hereditary information to the progeny, *Virology,* **33**:425–441, 1967.

Rosset, R., and Gorini, L., A ribosomal ambiguity mutation, *J. Mol. Biol.,* **39**:95–112, 1969.

Rottländer, E., Hermann, K. O., and Hertel, R., Increased heterozygote frequency in certain regions of the T4-chromosome, *Molec. Gen. Genetics,* **99**:34–39, 1967.

Rudkin, G. T., The structure and function of heterochromatin. *Proc. XI. Intl. Cong. Genetics,* **2**:359–374, 1963.

Rudner, R., Mutation as an error in base pairing, *Biochem. Biophys. Res. Commun.,* **3**:275–280, 1960.

Rudner, R., Mutation as an error in base pairing, 1. The mutagenicity of base analogues and their incorporation into the DNA of *Salmonella typhimurium,* *Z. Vererbungslehre,* **92**:336–360, 1961a.

Rudner, R., Mutation as an error in base pairing, 2. Kinetics of 5-bromodeoxyuridine and 2–aminopurine-induced mutagenesis, *Z. Vererbungslehre,* **92**:361–379, 1961b.

Rupert, C. S., Goodgal, S. H., and Herriott, R. M., Photoreactivation in vitro of ultraviolet inactivated *Hemophilus influenzae* transforming factor, *J. Gen. Physiol.,* **41**:451–471, 1958.

Rupp, W. D., and Howard-Flanders, P., Discontinuities in the DNA synthesized in an excision-defective strain of *Escherichia coli* following ultraviolet irradiation, *J. Mol. Biol.,* **31**:291–304, 1968.

Rutberg, B., A class of hybrids between bacteriophages T4B and T4D, *Virology,* **37**:243–251, 1969.

Ryan, F. J., Mutation in non-dividing bacteria, *Proc. Intl. Genetics Symp. (Tokyo-Kyoto),* 555–558, 1956.

Ryan, F. J., Nakada, D., and Schneider, M. J., Is NDA replication a necessary condition for spontaneous mutation? *Z. Vererbungslehre,* **92**:38–41, 1961.

Sambrook, J. F., Farr, D. P., and Brenner, S., A strong suppressor specific for UGA, *Nature,* **214**:452–453, 1967.

Sarabhai, A., and Brenner, S., A mutant which reinitiates the polypeptide chain after chain termination, *J. Mol. Biol.,* **27**:145–162, 1967.

Sarabhai, A. S., Stretton, A. O. W., Brenner, S., and Bolle, A., Co-linearity of the gene with the polypeptide chain, *Nature,* **200**:13–17, 1964.

Sauerbier, W., and Haug, A., An approach to the determination of the maximal contribution of thymine dimer to ultraviolet-inactivation of bacteriophage T4*vx, J. Mol. Biol.,* **10**:180–182, 1964.

Sauerbier, W., and Hirsch-Kauffmann, M., Transfer of ultraviolet light induced thymine dimer from parental to progeny DNA in bacteriophage T1 and T4, *Biochem. Biophys. Res. Commun.,*33:32–37, 1968.

Scafati, A. R., Streptomycin suppression of ambivalent phage mutations, *Virology,* **32**:543–552, 1967.

Scaife, J., F-prie factor formation in *E. coli* K12, *Genetical Res.,* 8:189–196, 1966.

Schlager, G., and Dickie, M. M., Spontaneous mutations and mutation rates in the house mouse, *Genetics,* 57:319–330, 1967.

Schuster, H., Die Reaktionsweise der Desoxyribonucleinsäure mit salpetriger Säure, *Z. Naturforsch.,* **15B**:298–304, 1960*a*.

Schuster, H., The reaction of nitrous acid with deoxyribonucleic acid, *Biochem. Biophys. Res. Commun.,* 2:320–323, 1960*b*.

Schuster, H., The reaction of tobacco mosaic virus ribonucleic acid with hydroxylamine,*J. Mol. Biol.,* 3:447–457, 1961.

Schuster, H., and Schramm, G., Bestimmung der biologisch wirksamen Einheit in der Ribosnucleinsäure des Tabakmosaikvirus auf chemischem Wege, *Z. Naturforsch.,* **13B**:697–704, 1958.

Schuster, H., and Vielmetter, W., Studies on the inactivating and mutagenic effect of nitrous acid and hydroxylamine on viruses, *J. Chim. Phys.,* **58**:1005–1010, 1961.

Schuster, H., and Wilhelm, R. C., Reaction differences between tobacco mosaic virus and its free ribonucleic acid with nitrous acid,*Biochim. Biophys. Acta,* **68**:554–560, 1963.

Schuster, H., and Wittmann, H. G., The inactivation and mutagenic action of hydroxylamine on tobacco mosaic virus ribonucleic acid, *Virology,* **19**:421–430, 1963.

Schwartz, N. M., Genetic instability in *Escherichia coli, J. Bacteriol,* **89**:712–717, 1965.

Scott, W. M., 5,6-Dihydro-6-methyluracil, a pyrimidine analogue more mutagenic than 5-bromouracil, *Bact. Proc.,* **55**, 1962.

Scotti, P. D., A new class of temperature conditional lethal mutants of bacteriophage T4D, *Mutation Res.,* 6:1–14, 1968.

Séchaud, J., Streisinger, G., Emrich, J., Newton, J., Lanford, H., Reinhold, H., and Stahl, M. M., Chromosome structure in phage T4, 2. Terminal redundancy and heterozygosis, *Proc. Natl. Acad. Sci. U.S.,* **54**:1333–1339, 1965.

Sellin, H. G., Srinivasan, P. R., and Borek, E., Studies of a phage-induced DNA methylase,*J. Mol. Biol.,* **19**:219–222, 1966.

Sesnowitz-Horn, S., and Adelberg, E. A., Proflavin treatment of *Escherichia coli*: generation of frameshift mutations, *Cold Spring Harbor Symp. Quant. Biol.,* 33:393–402, 1968.

Setlow, J. K., Evidence for a particular base alteration in two T4 bacteriophage mutants, *Nature,* **194**:664–666, 1962.

Setlow, J. K., The molecular basis of biological effects of ultraviolet radiation and photoreactivation, in *Current Topics in Radiation Research,* edited by M. Ebert and A. Howard, Vol. 2, pp. 195–248. North-Holland Publishing Co., Amsterdam, 1966.

Setlow, J. K., The effects of ultraviolet radiation and photoreactivation, in

Comprehensive biochemistry, edited by M. Florkin and E. H. Stotz, Vol. 27, pp. 157–209, Elsevier, New York, 1967.

Setlow, J. K., Boling, M. E., and Bollum, F. J., The chemical nature of photoreactivable lesions in DNA, *Proc. Natl. Acad. Sci. U.S.,* 53:1430–1436, 1965.

Setlow, J. K., and Setlow, R. B., Contribution of dimers containing cytosine to ultra-violet inactivation of transforming DNA, *Nature,* 213:907–909, 1967.

Setlow, R. B., and Carrier, W. L., The disappearance of thymine dimers from DNA: an error-correcting mechanism, *Proc. Natl. Acad. Sci. U.S.,* 51:226–231, 1964.

Setlow, R. B., and Carrier, W. L., The *v*-gene controls dimer excision in T4 phage, *Biophys. Soc. Abstracts,* 68, 1966.

Setlow, R. B., and Carrier, W. L., Formation and destruction of pyrimidine dimers in polynucleotides by ultra-violet irradiation in the presence of proflavin, *Nature,* 213:906–907, 1967.

Shalitin, C., and Stahl, F. W., Additional evidence for two kinds of heterozygotes in phage T4, *Proc. Natl. Acad. Sci. U.S.,* 54:1340–1341, 1965.

Shankel, D. M., and Kleinberg, J. A., Comparison of mutational synergism elicited by caffeine and acriflavin with ultraviolet light, *Genetics,* 56:589, 1967.

Shapiro, J. A., Mutations caused by the insertion of genetic material into the galactose operon of *Escherichia coli, J. Mol. Biol.,* 40:93–105, 1969.

Shapiro, R., and Klein, R. S., The deamination of cytidine and cytosine by acidic buffer solutions. Mutagenic implications, *Biochem.,* 5:2358–2362, 1966.

Shapiro, R., and Pohl, S. H., The reaction of ribonucleosides with nitrous acid. Side products and kinetics, *Biochem,* 7:448–455, 1968.

Shugar, D., and Fox, J. J., Spectrophotometric studies of nucleic acid derivatives and related compounds as a function of pH, *Biochim. Biophys. Acta,* 9:199–218, 1952.

Sicard, A. M., Mutagenèse induite par la proflavin chez le pneumocoque, *C. R. Acad. Sci.,* 261:4914–4917, 1965.

Siegel, A., Artificial production of mutants of tobacco mosaic virus, *Adv. Virus Res,* 11:25–60, 1965.

Siegel, A., Wildman, S. G., and Ginoza, W., Sensitivity to ultraviolet light of infective tobacco mosaic virus nucleic acid, *Nature,* 178:1117–1118, 1956.

Siegel, E. C., and Bryson, V., Selection of resistant strains of *Escherichia coli* by antibiotics and antibacterial agents: role of normal and mutator strains, *Antimicrobial Agents and Chemotherapy,* 1963:629–634, 1964.

Siegel, E. C., and Bryson, V., Mutator gene of *Escherichia coli* B. *J. Bacteriol.,* 94:38–47, 1967.

Signer, E. R., Beckwith, J. R., and Brenner, S., Mapping of suppressor loci in *Escherichia coli, J. Mol. Biol.,* 14:153–166, 1965.

Silver, S., Acriflavin resistance: a bacteriophage mutation affecting the uptake of dye by the infected bacterial cells, *Proc. Natl. Acad. Sci. U.S.,* 53:24–30, 1965.

Silver, S., Acridine sensitivity of bacteriophage T2: a virus gene affecting cell permeability, *J. Mol. Biol.,* 29:191–202, 1967.

Silver, S., Levine, E., and Pilelman, P. M., Acridine binding by *Escherichia coli*: pH dependency and strain differences, *J. Bacteriol.,* 95:333–339, 1968.

Simon, M. I., Grossman, L., and Van Vunakis, H., Photosensitized reaction of polyribonucleotides, 1. Effects on their susceptibility to enzyme digestion and their ability to act as synthetic messengers, *J. Mol. Biol.,* 12:50–59, 1965.

Simon, M. I., and Van Vunakis, H., The photodynamic reaction of methylene blue with deoxyribonucleic acid, *J. Mol. Biol.*, 4:488–499, 1962.

Simon, M. I., and Van Vunakis, H., The dye-sensitized photooxidation of purine and pyrimidine derivatives, *Arch. Biochem. Biophys.*, 105:197–206, 1964.

Singer, B., and Fraenkel-Conrat, H., Dye-catalyzed photoinactication of tobacco mosaic virus ribonucliec acid, *Biochemistry*, 5:2446–2450, 1966.

Singer, B., and Fraenkel-Conrat, H., Chemical modification of viral RNA, 6. The action of N-methyl-N′-nitro-N-nitrosoguanidine, *Proc. Natl. Acad. Sci. U.S.*, 58:234–239, 1967.

Singer, B.,Fraenkel-Conrat, H., Greenberg, J., and Michelson, A. M., Reaction of nitrosoguanidine (N-methyl-N′-nitro-N-nitrosoguanidine) with tobacco mosaic virus and its RNA, *Science,* 160:1235–1237, 1968.

Skaar, P. D., A binary mutability system in *Escherichia coli, Proc. Natl. Acad. Sci. U.S.*, 42:245–249, 1956.

Skalka, A., Burgi, E., and Hershey, A. D., Segmental distribution of nucleotides in the DNA of bacteriophage lambda, *J. Mol. Biol.*, 341:16, 1968.

Slapikoff, S., and Berg, P., Mechanism of ribonucleic acid polymerase action. Effect of nearest neighbors on competition between uridine triphosphate and uridine triphosphate analogues for incorporation into ribonucleic acids, *Biochem.*, 6:3654–3658, 1967.

Smith, K. C., The photochemical interaction of deoxyribonucleic acid and protein in vivo and its biological importance, *Photochem. Photobiol.*, 3:415–427, 1964.

Smith, K. C., Physical and chemical changes induced in nucleic acids by ultraviolet light, *Radiation Res.*, Suppl. 6:54–79, 1966.

Smith, K. C., and Aplin, R. T., A mixed photoproduct of uracil and cysteine (5-S-cysteine-6-hydrouracil). A possible model for the *in vivo* cross-linking of deoxyribonucleic acid and protein by ultraviolet light, *Biochem.*, 5:2125–2130, 1966.

Smith, K. C., Hodgkins, B., and O'Leary, M. E., The biological importance of ultraviolet light induced DNA-protein crosslinks in *Escherichia coli* 15 TAU, *Biochim. Biophys. Acta*, 114:1-15, 1966.

Söll, D., Studies on polynucleotides, 85. Partial purification of an amber suppressor tRNA and studies on *in vitro* suppression, *J. Mol. Biol.*, 34:175–187, 1968.

Söll, D., Jones, D. S., Ohtsuka, E., Faulkner, R. D., Lohrmann, R., Hayatsu, H., Khorana, H. G., Cherayil, J. D., Hempel, A., and Bock, R. M., Specificity of sRNA for recognition of codons as studied by the ribosomal binding technique, *J. Mol. Biol.*, 19:556–573, 1966.

Sompolinsky, D., Ben-Yakov, M., Aboud, M., and Boldur, I., Transferable resistance factors with mutator effect in *Salmonella typi, Mutation Res.*, 4:119–127, 1967.

Sonneborn, T. M., Gene and cytoplasm, 2. The bearing of the determination and inheritance of characters in *Paramecium aurelia* on the problems of cytoplasmic inheritance, Pneumococcus transformations, mutations and development, *Proc. Natl. Acad. Sci. U.S.*, 29:338–343, 1943.

Sonneborn, T. M., The differentiation of cells, *Proc. Natl. Acad. Sci. U.S.*, 51:915–929, 1964.

Speyer, J. F., Mutagenic DNA polymerase, *Biochem. Biophys. Res. Commun.*, 21:6–8, 1965.

Speyer, J. F., Base analogues and DNA polymerase mutagenesis, *Fed. Proc.*, 28:348, 1969.

Speyer, J. F., Karam, J. D., and Lenny, A. B., On the role of DNA polymerase in base selection, *Cold Spring Harbor Symp. Quant. Biol.,* 31:693–697, 1966.

Speyer, J., Karam, J., Rosenberg, D., and Lenny, A., Mutagenic polymerase in T4 phage, *Cold Spring Harbor Lab. Quant. Biol. Ann. Report,* 11–12, 1967.

Stadler, L. J., The gene, *Science,* 120:811–819, 1954.

Stafford, R. S., and Donnellan, J. E., Jr., Photochemical evidence for conformation changes in DNA during germination of bacterial spores, *Proc. Natl. Acad. Sci. U.S.,* 59:822–828, 1968.

Stahl, F. W., Edgar, R. S., and Steinberg, J., The linkage map of bacteriophage T4, *Genetics,* 50:539–552, 1964.

Stahl, F. W., Modersohn, H., Terzaghi, B. E., and Crasemann, J. M., The genetic structure of complementation heterozygotes, *Proc. Natl. Acad. Sci. U.S.,* 54:1342–1345, 1965.

Stahl, F. W., Murray, N. E., Nakata, A., and Crasemann, J. M., Intragenic *cis-trans* position effects in bacteriophage T4, *Genetics,* 54:223–232, 1966.

Steiger, H., and Kaplan, R. W., Induktion der Phagenproduktion und Auslösugn von Klarplaquemutationen durch UV-Bestrahlung des Prophagen und des freien Phagen Kappa bei Verwendung verschiedener *Serratia*-Stämme als Wirte, *Allg. Mikrobiol.,* 4:367–389, 1964.

Steinberg, C., and Stahl, F., The theory of formal phage crosses, *Cold Spring Harbor Symp. Quant. Biol.,* 23:42–45, 1958.

Steinberg, C., and Stahl, F., The clone-size distribution of mutants arising from a steady-state pool of vegetative phage, *J. Theoret. Biol.,* 1:488–497, 1961.

Stent, G. S., *Molecular Biology of Bacterial Viruses,* W. H. Freeman and Co., San Francisco, 1963.

Stent, G. S., Genetic transcription, *Proc. Roy. Soc. London,* (B), 164:181–197, 1966.

Stevenson, A. C., and Kerr, C. B., On the distribution of frequencies of mutation to genes determining harmful traits in man, *Mutation Res.,* 4:339–352, 1967.

Stewart, C. R., Mutagenesis by acridine yellow in *Bacillus subtilis, Genetics,* 59:23–31, 1968.

Stouthamer, A. H., DeHaan, P. G., and Bulten, E. J., Kinetics of F-curing by acridine orange in relation to the number of F-particles in *Escherichia coli, Genetical Res.,* 4:305–317, 1963.

Strack, H. B., Freese, E. B., and Freese, E., Comparison of mutation and inactivation rates induced in bacteriophage and transforming DNA by various mutagens, *Mutation Res.,* 1:10–21, 1964.

Strauss, B. S., Specificity of the mutagenic action of the alkylating agents, *Nature,* 191:730–731, 1961.

Strauss, B. S., Response of *Escherichia coli* auxotrophs to heat after treatment with mutagenic alkyl methanesulfonates, *J. Bacteriol.,* 83:241–249, 1962a.

Strauss, B. S., Differential destruction of the transforming activity of damaged DNA by a bacterial enzyme, *Proc. Natl. Acad. Sci. U.S.,* 48:1670–1675, 1962b.

Strauss, B. S., and Reiter, H., Repair of damage induced by a monofunctional alkylating agent in a transformable, ultraviolet-sensitive strain of *Bacillus subtilis, Genetics,* 52:478, 1965.

Streisinger, G., The genetic control of host range and serological specificity in bacteriophages T2 and T4, *Virology,* 2:377–387, 1956.

Streisinger, G., Edgar, R. S., and Denhardt, G. H., Chromosome structure in phage T4, 1. Circularity of the linkage map, *Proc. Natl. Acad. Sci. U.S.,* 51:775–779, 1964.

Streisinger, G., Emrich, J., and Stahl, M. M., Chromosome structure in phage T4, 3. Terminal redundancy and length determination, *Proc. Natl. Acad. Sci. U.S.,* **57**:292–295, 1967.

Streisinger, G., and Franklin, N. C., Mutation and recombination at the host range genetic region of phage T2, *Cold Spring Harbor Symp. Quant. Biol.,* **21**:103–111, 1956.

Streisinger, G., Mukai, F., Dreyer, W. J., Miller, B., and Harrar, G., Genetic studies concerning the lysozyme of phage T4, *J. Chim. Phys.,* **58**:1064–1067, 1961.

Streisinger, G., Okada, Y., Emrich, J., Newton, J., Tsugita, A., Terzaghi, E., and Inouye, M., Frameshift mutations and the genetic code, *Cold Spring Harbor Symp. Quant. Biol.,* **31**:77–84, 1966.

Streisinger, G., and Weigle, J., Properties of bacteriophages T2 and T4 with unusual inheritance, *Proc. Natl. Acad. Sci. U.S.,* **42**:504–510, 1956.

Strelzoff, E., Identification of base pairs involved in mutations induced by base analogues, *Biochem. Biophys. Res. Commun.,* **5**:384–388, 1961.

Strelzoff, E., DNA synthesis and induced mutations in the presence of 5-bromouracil, 2. Induction of mutations, *Z. Vererbungslehre,* **93**:301–318, 1962.

Strelzoff, E., and Ryan, F. J., The necessary involvement of both complementary strands of DNA in the specification of messenger RNA, *Biochem. Biophys. Res. Commun.,* **7**:471–476, 1962.

Stretton, A. O. W., Kaplan, S., and Brenner, S., Nonsense codons, *Cold Spring Harbor Symp. Quant. Biol.,* **31**:173–179, 1966.

Strigini, P., On the mechanism of spontaneous reversion and genetic recombination in bacteriophage T4, *Genetics,* **52**:759–776, 1965.

Sturtevant, A. H., *A History of Genetics,* Harper & Row, New York, 1965.

Sueoka, N., Correlation between base composition of deoxyribonucleic acid and amino acid composition of protein, *Proc. Natl. Acad. Sci. U.S.,* **47**:1141–1149, 1961.

Süsmuth, R., and Lingens, F., Zum pH-optimum der Mutationsauslösung durch 1-Nitroso-3-nitro-1-methyl-guanidin, *Naturwiss.,* **55**:85, 1968.

Sussenbach, J. S., and Berends, W., Photosensitized inactivation of deoxyribonucleic acid, *Biochim. Biophys. Acta,* **76**:154–156, 1963.

Sussenbach, J. S., and Berends, W., Photodynamic degradation of guanine, *Biochim. Biophys. Acta,* **95**:184–185, 1965.

Szybalski, W., Observations on chemical mutagenesis in microorganisms, *Ann. N. Y. Acad. Sci.,* **76**:475–489, 1958.

Takemoto, K. K., and Liebhaber, H., Virus-polysaccharide interactions, 1. An agar polysaccharide determining plaque morphology of EMC virus, *Virology,* **14**:456–462, 1961.

Tao, M., Small, G. D., and Gordon, M. P., Evidence for the participation of solvent in the ultraviolet inactivation of tobacco mosaic virus ribonucleic acid, *Fed. Proc.,* **26**:565, 1967.

Taylor, A. L., Bacteriophage-induced mutation in *Escherichia coli, Proc. Natl. Acad. Sci. U.S.,* **50**:1043–1051, 1963.

Taylor, H. F. W., The dissociation constants of benziminazole and certain purine derivatives, *J. Chem. Soc.,* 765–766, 1948.

Terzaghi, B. E., Streisinger, G., and Stahl, F. W., The mechanism of 5-bromouracil mutagenesis in the bacteriophage T4, *Proc. Natl. Acad. Sci. U.S.,* **48**:1519–1524, 1962.

Terzaghi, E., Okada, Y., Streisinger, G., Emrich, J., Inouye, M., and Tsugita, A.,

Change of a sequence of amino acids in phage T4 lysozyme by acridine-induced mutations, *Proc. Natl. Acad. Sci. U.S.,* **56**:500–507, 1966.

Tessman, E. S., Growth and mutation of phage T1 on ultraviolet-irradiated host cells, *Virology,* **2**:679–688, 1956.

Tessman, E. S., Mutants of bacteriophage S13 blocked in infectious DNA synthesis, *J. Mol. Biol.,* **17**:218–236, 1966.

Tessman, E. S., and Ozaki, T., The interaction of phage S13 with ultraviolet-irradiated host cells and properties of the ultraviolet-irradiated phage, *Virology,* **12**:431–449, 1960.

Tessman, E. S., and Shleser, R., Genetic recombination between phages S13 and ΦX174, *Virology,* **19**:239–240, 1963.

Tessman, E. S., and Tessman, I., Genetic recombination in phage S13, *Virology,* **7**:465–467, 1959.

Tessman, I., Mutagenesis in phages ΦX174 and T4 and properties of the genetic material, *Virology,* **9**:375–385, 1959.

Tessman, I., Mutagenesis and the functioning of genetic material in phage, in *The Molecular Basis of Neoplasia,* pp. 172–179, University of Texas Press, Austin, 1962*a*.

Tessman, I., The induction of large deletions by nitrous acid. *J. Mol. Biol.,* **5**:442–445, 1962*b*.

Tessman, I., Genetic ultrafine structure in the T4rII region, *Genetics,* **51**:63–75, 1965.

Tessman, I., Mutagenic treatment of double- and single-stranded DNA phages T4 and S13 with hydroxylamine, *Virology,* **35**:330–333, 1968.

Tessman, I., Ishiwa, H., and Kumar, S., Mutagenic effects of hydroxylamine in vivo, *Science,* **148**:507–508, 1965.

Tessman, I., Ishiwa, H., Kumar, S., and Baker, R., Bacteriophage S13: a seventh gene, *Science,* **156**:824–825, 1967.

Tessman, I., Kumar, S., and Tessman, E. S., Direction of translation in bacteriophage S13, *Science,* **158**:267–268, 1967.

Tessman, I., Poddar, R. K., and Kumar, S., Identification of the altered bases in mutated single-stranded DNA, 1. *In vitro* mutagenesis by hydroxylamine, ethyl methanesulfonate and nitrous acid, *J. Mol. Biol.,* **9**:352–363, 1964.

Thiry, L., Chemical mutagenesis of Newcastle disease virus, *Virology,* **19**:225–236, 1963.

Thomas, C. A., The arrangement of information in DNA molecules, *J. Gen. Physiol.,* **49** no. 6 part 2:143–169, 1966*a*.

Thomas, C. A., Recombination of DNA molecules, *Prog. Nucleic Acid Res. Mol. Biol.,* **5**:315–348, 1966*b*.

Tikchonenko, T. I., Dobrov, E. N., Velikodvorskaya, G. A., and Kisseleva, N. P., Peculiarities of the secondary structure of phage DNA *in situ, J. Mol. Biol.,* **18**:58–67, 1966.

Tocchini–Valentini, G. P., Stodolsky, M., Aurisicchio, A., Sarnat, M., Graziosi, F., Weiss, S. B., and Geiduschek, E. P., On the asymmetry of RNA synthesis *in vivo, Proc. Natl. Acad. Sci. U.S.,* **50**:935–941, 1963.

Tomizawa, J., and Ogawa, T., Inhibition of growth of rII mutants of bacteriophage T4 by immunity substance of bacteriophage lambda, *J. Mol. Biol.,* **23**:277–280, 1967.

Tomoeda, M., Inuzuka, M., Kubo, N., and Nakamura, S., Effective elimination of drug resistance and sex factors in *Escherichia coli* by sodium dodecyl sulfate, *J. Bacteriol.* **95**:1078–1089, 1968.

Trautner, T. A., Swartz, M. N., and Kornberg, A., Enzymatic synthesis of deoxyribonucleic acid, 10. Influence of bromouracil substitutions on replication, *Proc. Natl. Acad. Sci. U.S.*, **48**:449–455, 1962.

Treffers, H. P., Spinelli, V., and Belser, N. O., A factor (or mutator gene) influencing mutation rates in *Escherichia coli, Proc. Natl. Acad. Sci. U.S.*, **40**:1064–1071, 1954.

Tsugita, A., and Fraenkel-Conrat, H., The amino acid composition and C-terminal sequence of a chemically invoked mutant of TMV, *Proc. Natl. Acad. Sci. U.S.*, **46**:636–642, 1960.

Tsugita, A., and Inouye, M., Complete primary structure of phage lysozyme from *Escherichia coli* T4, *J. Mol. Biol.*, **37**:201–212, 1968.

Tsugita, A., Inouye, M., Imagawa, T., Nakanishi, T., Okada, Y., Emrich, J., and Streisinger, G., Frameshift mutations resulting in the changes of the same amino acid residue (140) in T4 bacteriophage lysozyme and *in vivo* codons for trp, tyr, met, val and ile, *J. Mol. Biol.*, **41**:349–364, 1969.

Tubbs, R. K., Ditmars, W. E., and VanWinkle, Q., Heterogeneity of the interaction of DNA with acriflavin, *J. Mol. Biol.*, **9**:545–557, 1964.

Ullmann, A., Jacob, F., and Monod, J., Characterization by *in vitro* complementation of a peptide corresponding to an operator-proximal segment of the β-galactosidase structural gene of *Escherichia coli, J. Mol. Biol.*, **24**:339–343, 1967.

Ullmann, A., Jacob, F., and Monod, J., On the subunit structure of wild-type versus complemented β-galactosidase of *Escherichia coli, J. Mol. Biol.*, **32**:1–13, 1968.

Valentine, R. C., and Zinder, N. D., Phenotypic repair of RNA-bacteriophage mutants by streptomycin, *Science*, **144**:1458–1459, 1964.

Van de Pol, J. H., and Van Arkel, G. A., The inactivating and mutagenic effect of hydroxylamine on bacteriophage ΦX174, *Mutation Res.*, **2**:466–469, 1965.

Van der Ent, G. M., Blok, J., and Linckens, E. M., The induction of mutations by ionizing radiation in bacteriophage ΦX174 and in its purified DNA, *Mutation Res.*, **2**:197–204, 1965.

Van Vunakis, H., Seaman, E., Kahan, L., Kappler, J. W., and Levine, Z., Formation of an adduct with tris(hydroxymethyl) aminomethane during the photooxidation of deoxyribonucleic acid and guanine derivatives, *Biochem.*, **5**:3986–3991, 1966.

Varghese, A. J., and Wang, S. Y., Thymine-thymine adduct as a photoproduct of thymine, *Science* **160**:186–187, 1968.

Vielmetter, W., and Schuster, H., Die Basenspezifität bei der Induktion von Mutationen durch salpetrige Säure im Phagen T2, *Z. Naturforsch.*, **15B**:304–311, 1960a.

Vielmetter, W., and Schuster, H., The base specificity of mutation induced by nitrous acid, *Biochem. Biophys. Res. Commun.*, **2**:324–328, 1960b.

Vielmetter, W., and Wieder, G. M., Mutagene und inaktivierende Wirkung salpetriger Säure auf freie Partikel des Phagen T2, *Z. Naturforsch.*, **14B**:312–317, 1959.

Visconti, N., and Delbruck, M., The mechanism of genetic recombination in phage, *Genetics*, **38**:5–33, 1953.

Vogt, M., Dulbecco, R., and Wenner, H. A., Mutants of poliomyelitis viruses with reduced efficiency of plating in acid medium and reduced neuropathogenicity, *Virology*, **4**:141–155, 1957.

Voll, M. J., Translation and polarity in the histidine operon, 3. The isolation of prototrophic polar mutations, *J. Mol. Biol.*, **30**:109–124, 1967.

von Borstel, R. C., Yeast genetics, *Science*, **152**:1287–1288, 1966.

von Borstel, R. C., On the origin of spontaneous mutations, *Proc. XII. Intl. Cong. Genetics*, 2:124, 1968.

von Borstel, R. C., Graham, D. E., La Brot, K. J., and Resnick, M. A., Mutator activity of a X-radiation-sensitive yeast, *Genetics*, 60:233, 1968.

Wacker, A., Molecular mechanisms of radiation effects, *Prog. Nucleic Acid Res.,* 1:369–399, 1963.

Wacker, A., Dellweg, H., Träger, L., Kornhauser, A., Lodemann, E., Türck, G., Selzer, R., Chandra, P., and Ishimoto, M., Organic photochemistry of nucleic acids, *Photochem. Photobiol.,* 3:369–394, 1964.

Wacker, A., Jacherts, J. D., and Jacherts, B., Effect of ultraviolet light on the biosynthesis of polyphenylalanine, *Angew. Chem. Intl. Edit.* 1:509, 1962.

Wacker, A., Kirschfeld, S., and Träger, L., Über den Einbau Purin-Analoger Verbindungen in die Bakterien-Nucleinsäure, *J. Mol. Biol.,* 2:241–242, 1960.

Wacker, A., Turck, G., and Gestenberger, A., Zum Wirkungsmechanismus photodynamischer Farbstoffe, *Naturwiss.,* 50:377, 1963.

Wang, S. Y., and Varghese, A. J., Cytosine-thymine addition product from DNA irradiated with ultraviolet light, *Biochem. Biophys. Res. Commun.,* 29:543–545, 1967.

Warner, H. R., and Barnes, J. E., Deoxyribonucleic acid synthesis in *Escherichia coli* infected with some deoxribonucleic acid polymerase-less mutants of bacteriophage T4, *Virology,* 28:100–107, 1966.

Watanabe, T., Transductional studies of thiamine and nicotinic acid requiring streptomycin resistant mutants of *Salmonella typhimurium, J. Gen. Microbiol.,* 22:102–112, 1960.

Watanabe, T., and Fukasawa, T., Episome-mediated transfer of drug resistance in *Enterobacteriaceae,* 2. Elimination of resistance factors with acridine dyes, *J. Bacteriol.,* 81:679–683, 1961.

Watson, J. D., The properties of X-ray inactivated bacteriophage, 1. Inactivation by direct effect, *J. Bacteriol.,* 60:697–718, 1950.

Watson, J. D., The properties of X-ray inactivated bacteriophage, 2. Inactivation by indirect effects, *J. Bacteriol.,* 63:473–485, 1952.

Watson, J. D., and Crick, F. H. C., The structure of DNA, *Cold Spring Harbor Symp. Quant. Biol.,* 18:123–131, 1953*a.*

Watson, J. D., and Crick, F. H. C., Genetical implications of the structure of deoxyribonucleic acid, *Nature,* 171:964–967, 1953*b.*

Weatherwax, R. S., Photoprotection of *Escherichia coli* from the lethal and mutagenic action of ultraviolet light, *Bact. Proc.,* 99, 1961.

Webb, R. B., and Kubitschek, H. E., Mutagenic and antimutagenic effects of acridine orange in *Escherichia coli, Biochem. Biophys. Res. Commun.,* 13:90–94, 1963.

Webb, R. B., and Malina, M. M., Mutagenesis in *Escherichia coli* by visible light, *Science,* 156:1104–1105, 1967.

Webb, R. B., Malina, M. M., and Benson, D. F., Action spectrum for mutagenesis by visible light in *Escherichia coli, Genetics,* 56:594–595, 1967.

Webb, R. B., and Petrusek, R. L., Oxygen effect in the protection of *E. coli* against U.V. inactivation and mutagenesis by acridine orange, *Photochem. Photobiol.,* 5:645–654, 1966.

Webber, B. B., and de Serres, F. J., Induction kinetics and genetic analysis of X-ray induced mutations in the *ad-3 region of Neurospora crassa, Proc. Natl. Acad. Sci. U.S.,* 53:430–437, 1965.

Weigert, M. G., Gallucci, E., Lanka, E., and Garen, A., Characteristics of the genetic code *in vivo, Cold Spring Harbor Symp. Quant. Biol.,* 31:145–150, 1966.

Weigert, M. G., and Garen, A., Base composition of nonsense codons in *E. coli*, *Nature*, **206**:992–994, 1965.

Weigle, J., Induction of mutations in a bacterial virus, *Proc. Natl. Acad. Sci. U.S.*, **39**:628–636, 1953.

Weigle, J. J., and Dulbecco, R., Induction of mutations in bacteriophage T3 by ultra-violet light, *Experientia*, **9**:372–373, 1953.

Weigle, J., Meselson, M., and Paigen, K., Modified density of transducing phage lambda, *Brookhaven Symp. Biol.*, **12**:125–133, 1959.

Weil, J., Terzaghi, B., and Crasemann, J., Partial diploidy in phage T4, *Genetics*, **52**:683–693, 1965.

Weill, G., and Calvin, M., Optical properties of chromophore-macromolecule complexes: absorption and fluorescence of acridine dyes bound to polyphosphates and DNA, *Biopolymers*, **1**:401–417, 1963.

Weinberg, R., and Boyer, H. W., Base analogue induced arabinose-negative mutants of *Escherichia coli*, *Genetics*, **51**:545–553, 1965.

Weisblum, B., Benzer, S., and Holley, R. W., A physical basis for degeneracy in the amino acid code, *Proc. Natl. Acad. Sci. U.S.*, **48**:1449–1454, 1962.

Weiss, B., and Richardson, C. C., Enzymatic breakage and joining of deoxyribonucleic acid, 1. Repair of single-strand breaks in DNA by an enzyme system from *Escherichia coli* infected with bacteriophage T4, *Proc. Natl. Acad. Sci. U.S.*, **57**:1021–1028, 1967.

Weiss, S. B., Hsu, W.-T., Foft, J. W., and Scherberg, N. H., Transfer RNA coded by the T4 bacteriophage genome, *Proc. Natl. Acad. Sci. U.S.*, **61**:114–121, 1968.

Welsh, J. N., and Adams, M. H., Photodynamic inactivation of bacteriophage, *J. Bacteriol.*, **68**:122–127, 1954.

Werner, R., Distribution of growing points in DNA of bacteriophage T4, *J. Mol. Biol.*, **33**:679–692, 1968.

Westmorland, B. C., Szybalski, W., and Ris, H., Mapping of deletions and substitutions in heteroduplex DNA molecules of bacteriophage lambda by electron microscopy, *Science*, **163**:1343–1348, 1969.

Whitehouse, H. L. K., and Hastings, P. J., The analysis of genetic recombination on the polaron hybrid DNA model, *Genetical Res.*, **6**:27–92, 1965.

Whitfield, H. J., Martin, R. G., and Ames, B. N., Classification of aminotransferase (*C* gene) mutants in the histidine operon, *J. Mol. Biol.*, **21**:335–355, 1966.

Wiemann, J., Zur Charakterisierung von Heterozygoten des Phagen T4, *Z. Vererbungslehre*, **97**:81–101, 1965.

Wilkins, M. F. H., Stokes, A. R., and Wilson, H. R., Molecular structure of deoxypentose nucleic acid, *Nature*, **171**:738, 1953.

Wills, C., Three kinds of genetic variability in yeast populations, *Proc. Natl. Acad. Sci. U.S.*, **61**:937–944, 1968.

Winkler, U., Studien über die uv-induzierte Mutabilität des *Serratia*-Phagen Kappa durch Versuche Mit uv-bestrahltem Indikator und Phagenkreuzungen, *Z. Naturforsch.*, **18B**:118–123, 1963.

Winkler, U., Wirtsreaktiverung von extrazellulär strahlen-induzierten Prämutationen im *Serratia*-Phagen *Kappa*, 1. Erhöhung der Mutabilität von *Kappa* durch Behinderung der Wirsreaktivierung, *Z. Vererbungslehre*, **97**:18–28, 1965*a*

Winkler, U., Wirtsreaktiverung von extrazellulär strahlen-induzierten Prämutationen im *Serratia*-phagen *Kappa*, 2. Über Pramutationen vollständig löschende Bakterienmutaten, *Z. Vererbungslehre*, **97**:29–39, 1965*b*.

Witkin, E. M., Mutations in *Escherichia coli* induced by chemical agents, *Cold Spring Harbor Symp. Quant. Biol.,* **12**:256–269, 1947.

Witkin, E. M., Modification of mutagenesis initiated by ultraviolet light through posttreatment of bacteria with basic dyes, *J. Cell. Comp. Physiol.,* **58** suppl. 1:135–144, 1961.

Witkin, E. M., The effect of acriflavin on photoreversal of lethal and mutagenic damage produced in bacteria by ultraviolet light, *Proc. Natl. Acad. Sci. U.S.,* **50**:425–430, 1963.

Witkin, E. M., Photoreversal and 'dark repair' of mutations to prototrophy induced by ultraviolet light in photoreactivable and non-photoreactivable strains of *Escherichia coli, Mutation Res.,* **1**:22–36, 1964.

Witkin, E. M., Mutation and the repair of radiation damage in bacteria, *Radiation Res.,* suppl. **6**:30–53, 1966.

Witkin, E. M., Mutation-proof and mutation-prone modes of survival in derivatives of *Escherichia coli* B differing in sensitivity to ultraviolet light, *Brookhaven Symp. Biol.,* **20**:17–53, 1968.

Witkin, E. M., Ultraviolet light-induced mutation and DNA repair, *Ann. Rev. Genetics,* **3**:(in press) 1969.

Witkin, E. M., and Sicurella, N. A., Pure clones of lactose-negative mutants obtained in *Escherichia coli* after treatment with 5-bromouracil, *J. Mol. Biol.,* **8**:610–613, 1964.

Witkin, E. M., Sicurella, N. A., and Bennett, G. M., Photoreversibility of induced mutations in a nonphotoreactivable strain of *Escherichia coli, Proc. Natl. Acad. Sci. U.S.,* **50**:1055–1059, 1963.

Witkin, E. M., and Theil, E. C., The effect of posttreatment with chloramphenicol on various ultraviolet-induced mutations in *Escherichia coli, Proc. Natl, Acad. Sci. U.S.,* **46**:226–231, 1960.

Wittmann, H. G., and Wittmann-Liebold, B., Protein chemical studies of two RNA viruses and their mutants, *Cold Spring Harbor Symp. Quant. Biol.,* **31**:163–172, 1966.

Woese, C. R., The present status of the genetic code, *Prog. Nucleic Acid Mol. Biol.,* **7**:107–172, 1967*a*.

Woese, C. R., *The Genetic Code: the Molecular Basis for Genetic Expression,* Harper and Row, New York, 1967*b*.

Wu, J.-H., Hildebrandt, A. C., and Riker, A. J., Virus-host relationships in plant tissue culture, *Phytopathology,* **50**:587–594, 1960.

Wulff, D. L., and Rupert, C. S., Disappearance of thymine photodimer in ultraviolet irradiated DNA upon treatment with a photoreactivating enzyme from baker's yeast, *Biochem. Biophys. Res. Commun.,* **7**:237–240, 1962.

Yamane, T., Wyluda, B. J., and Shulman, R. G., Dihydrothymine from UV-irradiated DNA, *Proc. Natl. Acad. Sci. U.S.,* **58**:439–442, 1967.

Yanofsky, C., Carlton, B. C., Guest, J. R., Helinski, D. R., and Henning, U., On the colinearity of gene structure and protein structure, *Proc. Natl. Acad. Sci. U.S.,* **51**:266–272, 1964.

Yanofsky, C., Cox, E. C., and Horn, V., The unusual mutagenic specificity of an *E. coli* mutator gene, *Proc. Natl. Acad. Sci. U.S.,* **55**:274–281, 1966.

Yanofsky, C., Drapeau, G. R., Guest, J. R., and Carlton, B. C., The complete amino acid sequence of the tryptophan synthetase A protein (α subunit) and colinear relationship with the genetic map of the A gene, *Proc. Natl. Acad. Sci. U.S.,* **57**:296–298, 1967.

Yanofsky, C., Horn, V., and Thorpe, D., Protein structure relationships revealed by mutational analysis, *Science,* **146**:1593–1594, 1964.

Yanofsky, C., and Ito, J., Nonsense codons and polarity in the tryptophan operon, *J. Mol. Biol.,* 21:313–334, 1966.

Yanofsky, C., and Ito, J., Orientation of antipolarity in the tryptophan operon of *Escherichia coli, J. Mol. Biol.,* 24:143–145, 1967.

Yanofsky, C., Ito, J., and Horn, V., Amino acid replacements and the genetic code, *Cold Spring Harbor Symp. Quant. Biol.,* 31:151–162, 1966.

Yoshikawa, H., Mutations resulting from the transformation of *Bacillus subtilis, Genetics,* 54:1201–1214, 1966.

Zakharov, I. A., Kozhina, T. N., and Fedorova, I. V., Increase in spontaneous mutability in U.V.-sensitive yeast mutants, *Dokl. Akad. USSR,* 181:470–472, 1968.

Zamenhof, P. J., A genetic locus responsible for generalized high mutability in *Escherichia coli, Proc. Natl. Acad. Sci. U.S.,* 56:845–852, 1966.

Zamenhof, S.,Effects of heating dry bacteria and spores on their phenotype and genotype, *Proc. Natl. Acad. Sci. U.S.,* 46:101–105, 1960.

Zamenhof, S., Nucleic acids and mutability, *Prog. Nucleic Acid Res. Mol. Biol.,* 6:1–38, 1967.

Zamenhof, S., Eichhorn, H. H., and Rosenbaum-Oliver, D., Mutability of stored spores of *Bacillus Subtilis, Nature,* 220:818–819, 1968.

Zamenhof, S., and Greer, S., Heat as an agent producing high frequency of mutations and unstable genes in *Escherichia coli, Nature,* 182:611–613, 1958.

Zampieri, A.,and Greenberg, J., Mutagenesis by acridine orange and proflavin in *Escherichia coli* strain S, *Mutation Res.,* 2:552–556, 1965.

Zampieri, A., Greenberg, J., and Warren, G., Inactivating and mutagenic effects of 1-methyl-3-nitro-1-nitrosoguanidine on intracellular bacteriophage. *J. Virology,* 2:901–904, 1968.

Zelle, M., Ogg, J. E., and Hollaender, A., Photoreactivation of induced mutation and inactivation of *Escherichia coli* exposed to various wavelengths of monochromatic ultraviolet radiation, *J. Bacteriol.,* 75:190–198, 1958.

Zinder, N. D., Infective heredity in bacteria, *Cold Spring Harbor Symp. Quant. Biol.,* 18:261–269, 1953.

Zipser, D., Orientation of nonsense codons on the genetic map of the *lac* operon, *Science,* 157:1176–1177, 1967.

Zipser, D., and Newton, A., The influence of deletions on polarity, *J. Mol. Biol.,* 25:567–569, 1967.

Author index

Subject index